Learning and Analytics in Intelligent Systems

Volume 57

The main aim of the series is to make available a publication of books in hard copy form and soft copy form on all aspects of learning, analytics and advanced intelligent systems and related technologies. The mentioned disciplines are strongly related and complement one another significantly. Thus, the series encourages cross-fertilization highlighting research and knowledge of common interest. The series allows a unified/integrated approach to themes and topics in these scientific disciplines which will result in significant cross-fertilization and research dissemination. To maximize dissemination of research results and knowledge in these disciplines, the series publishes edited books, monographs, handbooks, textbooks and conference proceedings.

Indexed by EI Compendex.

Kostas Karpouzis · Yannis Skarpelos
Editors

Computational Methods for the Digital Humanities

Springer

Editors
Kostas Karpouzis
Department of Communication, Media and Culture
Panteion University of Social and Political Sciences
Athens, Greece

Yannis Skarpelos
Department of Communication, Media and Culture
Panteion University of Social and Political Sciences
Athens, Greece

ISSN 2662-3447 ISSN 2662-3455 (electronic)
Learning and Analytics in Intelligent Systems
ISBN 978-3-032-05098-4 ISBN 978-3-032-05099-1 (eBook)
https://doi.org/10.1007/978-3-032-05099-1

This Springer imprint is published by the registered company Springer Nature Switzerland AG
The registered company address is: Gewerbestrasse 11, 6330 Cham, Switzerland

Introduction

The rapid advancement of digital technologies has significantly altered the study of the humanities, introducing both opportunities and challenges to understanding culture, history, and society. Computational methods have grown into powerful tools, enabling researchers to process, analyze, and interpret vast datasets with an unprecedented level of precision. These methods allow us to uncover patterns and connections that were previously hidden, thus expanding the scope and depth of humanistic research. This book is titled *Computational Methods for the Digital Humanities* and provides a comprehensive examination of how computational tools intersect with humanistic inquiry, discussing theoretical frameworks as well as practical applications.

Digital humanities is a multidisciplinary field that integrates insights from computer science, social sciences, and traditional humanities disciplines. It spans a broad array of methodologies, ranging from natural language processing and data visualization to critical examinations of the biases inherent in computational tools. As such, this field presents a unique combination of possibilities and challenges. The chapters in this volume collectively represent this diversity, demonstrating how computational methods can illuminate cultural and social phenomena while also critiquing the underlying systems and structures that frame these methods. By applying computational techniques to humanistic questions, this field not only advances scholarly inquiry but also raises profound questions about ethics, inclusivity, and the future of research.

This book is written for scholars, practitioners, and students eager to understand how computational approaches can transform humanistic research. Each chapter offers a unique perspective, presenting case studies, methodologies, and critical analyses that highlight the potential of computational methods to reshape the way we study the humanities. At the same time, the volume emphasizes the importance of adopting ethical practices and fostering inclusivity to ensure that digital humanities remains a space for diverse voices and perspectives. The goal is not only to showcase technological innovation but also to critically engage with its broader implications.

The chapters are arranged thematically, providing a structured exploration of key areas within computational methods for the digital humanities. In Chap. 1,

Yannis Skarpelos and Sophia Messini examine the role of mass audiences in curating and sharing images on social media. By analyzing approximately 23,000 images related to the announcement of the COVID-19 Omicron variant, the authors explore the cultural and emotional factors that drive image selection. Their findings highlight the interplay between textual content, visual imagery, and emotional expression, shedding light on the dynamics of online communication.

Chapter 2 explores the use of digital storytelling platforms to preserve and interpret cultural heritage. By focusing on first-person narratives, author Foteini Venieri reveals how computational tools can create immersive experiences that connect audiences with historical and cultural contexts. The chapter underscores the value of integrating personal perspectives into heritage narratives, offering new ways to communicate and understand history.

In Chap. 3, Myrto Koukouli and colleagues present *CoVEd*, a user-friendly editor designed for creating and managing controlled vocabularies within the SKOS schema. The chapter discusses the challenges faced by cultural heritage professionals in adopting linked data practices and demonstrates how CoVEd simplifies these processes. By making data interoperability more accessible, this tool supports more effective knowledge sharing and collaboration.

Chapter 4 critically investigates the intersection of artificial intelligence and digital humanities through the lens of colonial power dynamics. Using case studies, Kostas Karpouzis illustrates how AI tools can perpetuate colonial biases while also offering possibilities for challenging them. Recommendations for inclusive AI design and community-driven data governance are provided, emphasizing the need to integrate diverse epistemologies in knowledge production.

Elpida Charitou, in Chap. 5, introduces a virtual museum created on the FrameVR platform to promote space science education. The chapter details the design and curation process, demonstrating how immersive technologies can make scientific concepts more accessible and engaging. Orbitorium exemplifies the potential of digital museums to democratize knowledge and enhance public understanding of complex topics.

The book concludes with two chapters focusing on applications of computational methods. First, Nikolaos Tantaroudas and colleagues explore the application of eye-tracking technology in extended reality environments to study consumer behavior. Chapter 6 highlights how this approach can provide insights into food choices and preferences, revealing biases and patterns that traditional methods might overlook. The findings have implications for research, marketing, and education, demonstrating the value of immersive technologies in behavioral analysis. Then, Thanasis Troboukis and collaborators examine the innovative application of hybrid intelligence, i.e., human collaboration with AI, in analyzing political speeches from the 2023 Greek general elections. Chapter 7 highlights the use of a Large Language Model (LLM) and interdisciplinary methods to assess sentiment, polarization, populism, and topics within campaign speeches, underscoring the potential and limitations of AI in journalism, and emphasizing the need for human oversight and ethical considerations.

Together, these diverse contributions illustrate the transformative potential of computational approaches for advancing research in the Social Sciences and Humanities. They also confront the ethical and methodological complexities inherent in such work. By integrating computational methods with critical perspectives, we strongly believe that the *Computational Methods for the Digital Humanities* book offers a multifaceted examination of how technology and humanistic inquiry intersect to produce innovative forms of knowledge and understanding.

Athens, Greece

Kostas Karpouzis
kkarpou@panteion.gr
https://culturalinformaticslab.gr/en/

Yannis Skarpelos
gskarp@panteion.gr
https://culturalinformaticslab.gr/en/

Contents

Chapter 1
Cultural and Emotional Grounds for Image Choice in Twitter: The Case of #omicron Variant

Yannis Skarpelos and Sophia Messini

Abstract Computational methods challenge conventional visual analysis tools, inviting to a deeper understanding of uses of visual images in massive volumes by audiences themselves. Audiences have always used found images or amateur photographs, but such practices intensified with the advent of Web 2.0 and social media. More and more they are understood as a kind of self-motivated, amateur curatorial work. In this paper we explore the cultural roots and emotional motivation behind image selection by mass audiences in social media. We take an instance in the COVID-19 pandemic as a case study: images used in Twitter the day Omicron variant was announced. Using computational methods, we collected and analysed approx.23,000 images. We found a divide between textual and pictorial images, which may be explained based on cultural grounds. We also found that different kinds of imagery were related to tweets expressing differences in certain emotions like Fear or Sadness, and on secondary emotions like Despair, Guilt and Aggressiveness. The findings show that the connection between the visual, the textual, and the emotional elements is a fruitful field for novel knowledge.

Keywords Social media · Visual semiotics · Sentiment analysis · COVID-19 · Computational analysis

1.1 Introduction

The advent of big data, the rise of digital humanities into prominence, the accumulation of visual corpora and their daily reconstruction with millions of images added to this universal Archive, all dictate a shift of paradigm in visual semiotics so far understood as the 'close reading' of small corpora [17]. In fact, such a shift is already

Y. Skarpelos (✉) · S. Messini
Department of Communication, Media and Culture, CID-CCS Lab,
Panteion University of Social and Political Sciences, Athens, Greece
e-mail: gskarp@panteion.gr
URL: https://culturalinformaticslab.gr/en/

K. Karpouzis and Y. Skarpelos (eds.), *Computational Methods for the Digital Humanities*,
Learning and Analytics in Intelligent Systems 57,
https://doi.org/10.1007/978-3-032-05099-1_1

under way, though still marginal. Replaced by 'distant reading' of big visual corpora dictates the use of computational methods in semiotics. Computational semiotics is "studying semiotic phenomena in a human-created content–all kinds of content, from text to artwork–with computational models and methods" [13]. Despite the idea of mathematical modelling of semiosis is still disputed, it is also challenging our traditional methods along with the conceptual framework within which to conduct semiotic analysis.

The Visual Universe is ever expanding exponentially, in a way like the Big Bang. Even if we narrow down the Visual to the photographic, we are left with an amount estimated close to 14.3 trillion increasing by 7% per year [11]. Of the 1.94 trillion photos estimated to be taken in 2024, a significant part will be shared online in social media platforms like WhatsApp, Snapchat, Facebook, Instagram, Flickr or Twitter.

Researchers have access to unprecedented amounts of visual data. Such data include both legacy images (e.g., paintings from all major museums) or newly created images (e.g., photos uploaded to social media around a specific theme or hashtag), while collections of archived artistic or photographic visuals are readily available by most libraries, galleries, or museums online. Within the occularcentric paradigm" [23, 26], such collections are considered as evidence about societies, social values, personal and social identities, and so on. They not only depict their social reality but also represent its hidden structure. Hamilton suggested that "Like all forms of photographic representation it is not simply a 'record' of a given moment, for it cannot be innocent of the values and ethics of those who worked within it" [21]. While Hamilton is not concerned with amateur photographic practice (which he anyway considers as influenced by the style of imagery associated with professional photography), visual scholars turned to it as a "rule-governed and socially patterned communicative process" [40, 46]. And beyond photographic practices, a wave of image appropriation is underway, with users oversignifying found visual materials with new meanings, with deviant connotations that subvert the original ones. In this, we are reminded of what Stuart Hall described as 'negotiated reading.'

Social media not only opened such processes to a wider audience but provoked "new image-making practices that are more attentive to context, social relations and materiality, and hence to the world surrounding the frame" [20]. Along with mainstream images they host unconventional imagery, like images that might be thought of as ugly [24], and transform the visual image to "become less of a static object produced and controlled by a sole author, but rather a transformative tool in the hands of millions of producers and perceivers" [5], social media platforms push visual culture to its edge, disseminate visual imagery in unprecedented degree, offer visibility to visual subcultures which otherwise would pass unacknowledged. On the other hand, it is algorithms that decide what is accessible to each user. Algorithmic curation of content results in social feed or timeline "mediates users' exposure to information deemed to be sufficiently relevant or interesting" [7]. Recent audits have shown that there are multiple biases in what a user actually sees in her timeline, such as popularity and exposure bias.

Moving from a single or a few images to a visual corpus signals distancing and the search for patterns, for regularities and irregularities beyond the visually perceptible.

Techniques used so far not only provide ways towards such patterns, but also open new possibilities to strengthen analytical rigor.

In this paper we undertake a double task: (a) we analyze the case of visual content uploaded to Twitter during the day when the variant omicron of COVID-19 was announced, and (b) we showcase the application of a computational tool set for visual analysis along with the methodological considerations behind its application. Those two tasks are completed in tandem. We present the steps taken together with the expose of our findings. This may seem unconventional but is an appropriate way to discuss new tools along with their application, showcasing both their potential and pitfalls.

1.2 Case Study: COVID-19, Tweets and Images

Computer scientists have been working on parallel analysis of texts and images in social networks (Twitter, Facebook, Instagram, Weibo etc.), aiming to automate processes distinguishing fake content. Similar works with reference to the COVID-19 pandemic are rather scarce, though the promise of such a parallel analysis might have implications both in evaluating the success of preventive measures taken by government, in supporting successful health-related communication campaigns, support people in isolation during movement restriction periods etc. Cheema et al. [14], compared the efficiency of different Machine Learning approaches to approaching the role of images in claim detection, which might prove useful in detecting fake or conspiratorial content. Negri et al used images posted in social media as a proxy to "policy-adherence with regards to mask wearing, social distancing, and other hard-to-measure quantities" [42], introducing "a framework for image-based social sensing." Tallal Javed et al. [22] compared the use, typology and lifetime of images and texts in WhatsApp and Twitter within the national framework of Pakistan. In a later paper, they reviewed the same content in relation to the political orientation of the users. Stella et al. [55] combined natural language processing, cognitive network science and AI-based image analysis on mask wearing, to study the semantic and emotional framing of COVID vaccine related tweets. Unal et al. [57] computationally analyse text and images in COVID related tweets to identify elements predictive to information popularity and reliability, aiming to uncover how unreliable articles construct multi-modal meaning in a distorted, biased fashion.

On a more communication-centred direction, Renshaw et al. [51] examined dimensions of textual and visual content leading to higher retransmission rate of official Tweets during the pandemic, while Sleigh et al. [54] content analysed most retweeted COVID related tweets to find that though 'health gain' visual messages spread more, most tweets frame 'health loss.' Pérez-Martínez [47] provided a review of tweets related to the Catholic Church during the first year of the pandemic, focusing on the *Urbi et Orbi* prayer by the Pope and its textual and visual reception in Twitter, along its sentimental evaluation. Bardakis [6] and Ponton [50] both handpicked memes from social media to analyze the workings of humor and satire during dire periods.

In our research we examine the relationship between the features of the images and the content of the text accompanying them. Are there distinctive patterns suggesting the existence of cultural traits rather than personal preferences as the *raison d'être* of image selection. Does image similarity correspond to similarity in sentiment and emotions expressed? Thus, are images with similar features in the plane of expression or in the plane of content related to texts expressing similar ideas, sentiment, or emotions? Thus, we first explore the 'fingerprints' of the images and provide a 'distant viewing' explanation, and then we turn to the texts to analyse the sentiments and emotions they express. Finally, we co-relate those sentiments and emotions to the images themselves.

1.3 Data and Methods

To achieve such a complex task, we used a series of computational methods and tools. A detailed presentation of the tools is provided in the Technical Appendix. A graphic presentation of the workflow is presented in Fig. 1.1.

Our corpus consists of all publicly available tweets posted during the day the variant Omicron of COVID-19 was announced (27-11-2021). The tweets were collected using an Academic Research license for accessing Twitter API, with twarc2 [56] command line tool for collecting and archiving Twitter JSON data via the Twitter API, on 6-6-2022. We searched for '#omicron' and found 807,767 tweets and retweets. The corpus consists of 586,674 tweets and 176,118 retweets, while the rest were not publicly available. The tweets were encoded as using 61 languages plus undefined. The languages present in the corpus with more than 1,000 tweets are presented in Table 1.1.

Since the search was conducted with the hashtag #omicron, languages with non-latin alphabets may be under-represented in our corpus. Thus, the corpus may lean

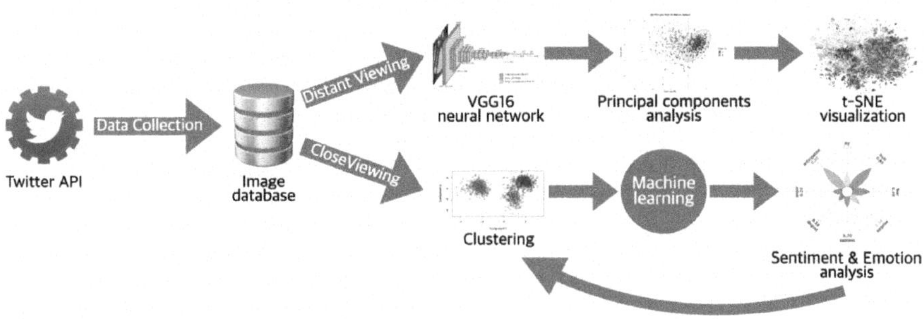

Fig. 1.1 The workflow used in this paper

Table 1.1 Languages included in the corpus

Language code	Language	Tweets	Percent (%)
en	English	444,157	58,23
es	Spanish, Castilian	95,699	12,55
th	Thai	55,709	7,30
fr	French	37,033	4,85
de	German	22,003	2,88
pt	Portuguese	20,068	2,63
it	Italian	14,946	1,96
und	Undefined	13,618	1,79
tr	Turkish	12,312	1,61
in	???	10,524	1,38
hi	Hindi	7,500	0,98
nl	Dutch, Flemish	5,942	0,78
ja	Japanese	5,180	0,68
zh	Chinese	2,795	0,37
ar	Arabic	1,598	0,21
pl	Polish	1,585	0,21
ca	Catalan, Valencian	1,350	0,18
el	Greek, Modern	1,260	0,17
	Other Language	9,513	1,25

towards occidentalism and be West-centric. Being alarmed at this stage is sufficient, since new case studies may challenge our findings and complete the corpus with new searches based on local variations and alphabets.

There were 56,290 tweets including images of which 22,952 were the originals. In some cases, identical or similar images were used from diverse sources, therefore they were present in our corpus with different names. We decided to keep them in the final visual corpus, following a similar strategy by Manovich et al. [33], who kept images from their dataset as many times as the hashtags they were related with, recognizing that this meant more possibilities for being viewed by diverse audiences. In a similar manner, an image disseminated by a photo agency to media worldwide, had the opportunity to be reproduced in multiple unrelated tweets thus reaching diverse national or other audiences.

We followed a two-fold methodology, taking advantage of an interplay between zooming in and out to our visual corpus: a macro-analytical 'distant viewing' methodology was utilized to achieve an overview of the data and analyse evident differences within our visual corpus (Figs. 1.2 and 1.3), followed by a 'close reading' of certain image clusters (Figs. 1.4, 1.5, 1.6 and 1.7).

1.3.1 Distant Viewing

Some researchers employ 'distant viewing' or "macroanalysis of images", understood as a tool for "making explicit the interpretive nature of extracting semantic metadata from images" [3]. The idea is familiar to Digital Humanities [58]. It was introduced by Franco Moretti in [38] and [37] as a means to deal with 'the great unread,' or the literary works that lie beyond the Western Canon. The challenge posed by Moretti was "a change in how we look at all of literary history: canonical and noncanonical: together" [38]. Manovich describes such an idea as driving his aspirations back in 2007: "The user (…) would switch among all three data views, zooming in and out, seeing trends in a larger context and making comparisons" [32].

While traditional methods in art history and visual studies celebrate the exception (i.e. "the 'leading' and 'most importantly' the 'avant-garde' cases") as canon [31], such techniques allow the researcher to access the opposite kind of outliers: the "invisible minorities, which in a normal sample are small enough to be considered as statistical error" [53, 61]. According to Dondero, "[t]he advantage of visualizing the entire collection and the structure of its component parts is that it combines a distant overview of the collection with the insights of every artifact" [17].

Arnold and Tilton describe this method as "automatic extraction of semantic elements of visual materials followed by the aggregation and visualization of these elements via techniques from exploratory data analysis" [3]. Its strength lies in the possibility to control the researcher's distance from the corpus: distant viewing does not rule out the possibility of close reading. Instead, the researcher may decide to zoom in to a sub-corpus, or even closer to a single image, and then zoom out again, to realize the patterns that connect it to the whole.

The interplay between distant and close reading is eventually an exercise in pattern recognition, either within images or among them. Pattern recognition tasks are completed by software, while the researcher decides upon visualization parameters. Then, it is up to the researcher to contemplate this new type of meta-image, discern the patterns and "make hypotheses about trends" in the visual corpus [17].

We used unsupervised techniques based on the VGG16 pre-trained convolutional network [52] to extract plastic features from the images and applied Principal Components Analysis (PCA) to reduce the dimensions of the features extracted from 4,096 to 43. We measured the distances between the features of each image to the rest of the corpus and visualised the result in a 2D space using t-distributed stochastic neighbor embedding (t-SNE) [29] as a media visualization, i.e., a scatterplot incorporating the actual images (Fig. 1.2). Finally, to achieve a conventional visualization, we used RasterFairy library for Python, which aims "to transform any kind of 2D point cloud into a regular raster whilst trying to preserve the neighborhood relations that were present in the original cloud" [27] (Fig. 1.3).

Fig. 1.2 Media visualization scatterplot of the complete image corpus

1.3.2 Close Viewing

In any case, 'big' is not just about the volume of visual data available and their visualization. The methodologies applied to such big corpora may be used in quite small corpora as well [9, 53, 61]. It is the methodologies and the frame of mind that impose a new way of seeing both data and images sociologically and semiotically.

Dondero points out that "*from a semiotic point of view, we can see and understand whether images that share the same characteristics on the plane of expression are more or less homogeneous on the plane of content and whether they are consistently generated by the same issues, studies and experiments. In this sense, quantitative analysis of visual categories on the plane of expression is preparatory to qualitative signification analysis.*" [17].

To 'see' closer the images in our corpus and relate them to certain emotions and feelings we had to limit them to those used in English language tweets. Since 'pictorial' images formed the larger part of the imagery collected in our corpus, we decided to further examine the divisions within this group. For convenience we used the open-source software Orange [16] Image Analytics addon, applied VGG16 pre-trained convolutional network image embeddings, computed distances, and conducted hierarchical clustering which led to four clusters.

Fig. 1.3 Media visualization of the complete image corpus

The first cluster, consisting of 3,429 images, was populated mostly with images taken from comics or cartoons. Within this cluster, a considerable part was dedicated to characters from the *Futurama* animated series, especially the villain *Lrrr* originating from planet Omicron Persei 8. Though the series ended in 2013, its audience was quick to adopt it as a meme. Images of the letter O (omicron) were also included in this cluster, along with emoticons and other memes (Fig. 1.4).

The second cluster is the smaller one, consisting of 955 images. It was populated with conventional graphic representations of coronavirus, in diverse versions and colours (Fig. 1.5).

The third cluster, consisting of 3,185 images, included in the most part infographics, clippings from online media and photos of newspapers, tables, and diagrams. In this cluster were placed images with the Greek alphabet which were used by conspiracy advocates to support the idea that skipping one letter (the letter Xi) was

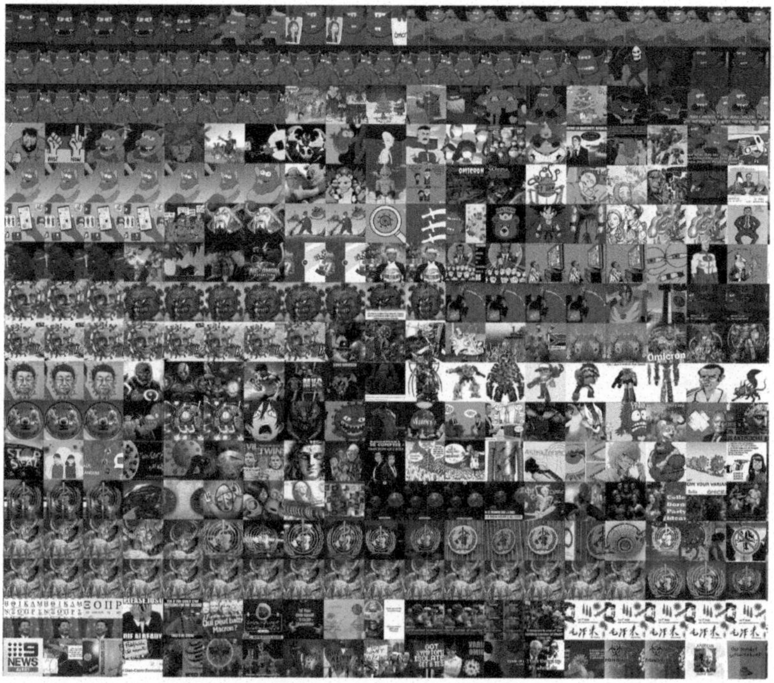

Fig. 1.4 The 'comics/cartoon' cluster (detail)

intentional in order to hide from public conscience the association with the President of China, Xi Jinping (Fig. 1.6).

The last group was the bigger one, consisting of 9,401 images. The majority were portraits: mostly portraits of politicians and eminent persons related to the management of pandemics, as well as journalists and ordinary citizens (Fig. 1.7).

1.3.3 Scrutinizing Emotions

Then we related each cluster with the sentiment, emotions and feelings expressed in the relevant tweets. There is already published research on the emotions expressed during the pandemic[1, 4, 34, 35] indicating a dynamic interaction between emotions and the social, political and medical environment and the measures taken to confront the pandemic in each country.

We applied Plutchik's theory of emotions to the tweet texts. Plutchik developed a psycho-biological approach to emotions, beyond the dichotomy between positive and negative sentiment. In short, he proposed a model "assuming that there are eight basic emotions which play major adaptive roles in both animals and humans (…) [which] can be combined in a large number of ways to produce the many derivative

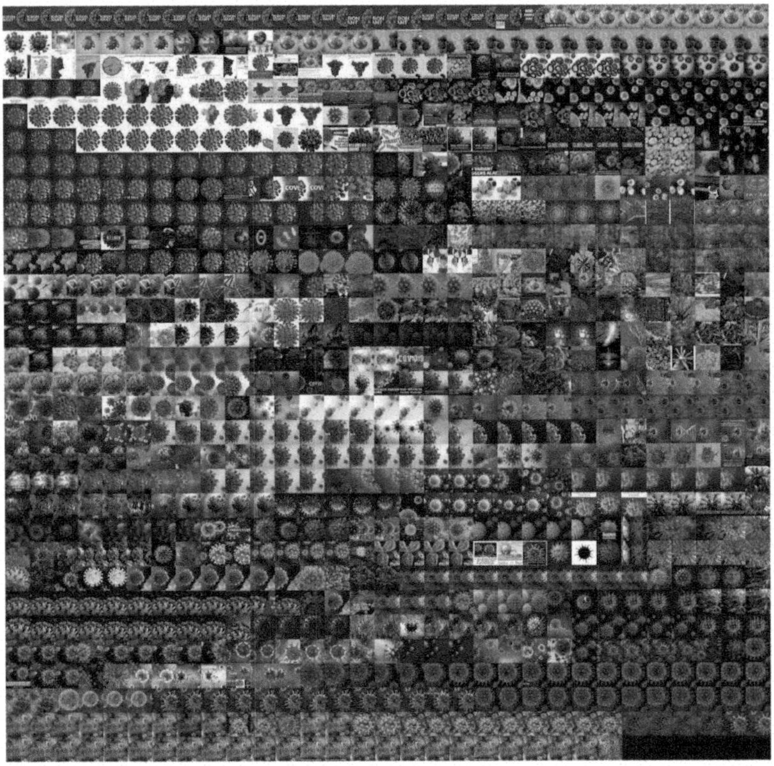

Fig. 1.5 The 'virus' cluster, containing mostly graphic representations of coronavirus

states which are actually observed" [48]. Originally developed in late 1950s, this theory was being elaborated and the idea of measuring emotions was examined in a volume at the end of 1980s [49], giving rise to the idea of its operationalization in computational linguistics. His model has been used recently in social science to evaluate their influence upon passengers' intention to recommend airlines during COVID-19 pandemic [60], the motivation behind 'vaccine tourism' (Gulati 2022), or the emotions expressed in Twitter *vis-à-vis* the #metoo movement [41], among others.

To apply Plutchik's theory of emotions to our corpus, we used the NRC Word-Emotion Association Lexicon or EmoLex created by Saif M. Mohammad [36]. The lexicon contains the emotional weights for 14,182 words, measuring negative and positive sentiment, along with eight emotions: anger, anticipation, disgust, fear, joy, sadness, surprise, trust.

Fig. 1.6 The 'infographics' cluster, containing also clippings from online media and photos of newspapers, tables, and diagrams (detail)

Fig. 1.7 The 'portraits' cluster (detail)

1.4 Findings

1.4.1 The 'textual' Versus 'pictorial' Images Divide

Figures 1.2 and 1.3 highlight a divide between the images placed on the upper left and those at the lower right part of the diagrams. The former are mostly images displaying text or some simple diagram followed by text, which we will call 'textual images,' and consists of almost one third (28%) of the corpus. The latter consists of pictures of any sort, which we will call 'pictorial images.' This divide is critical in understanding media use strategies of social media users, and better evaluate the relationship between the tweet itself and its visual elements. While a few studies have been discussing such a relationship, the mostly focus on the degree of similarity between the two, their semantic overlap, or their relevance [39, 59]. Moreover, such strategies may pave the way to rhetorical devices like irony, or sarcasm [12, 62]. They may also be related to social, ideological, and cultural characteristics of the users.

We explored the 'textual' versus 'pictorial' images reading 1,000 randomly selected tweet texts from each category. For the tweets in languages we were not familiar with, we used Google Translate or Deepl.com. We realized that there was no evidence that they were related to known divisions as the sceptics, the antivax movement or conspiracy theories. In fact, conspirationists were mostly interested in the skipped letter which in the Greek alphabet adopted so far to indicate COVID variants should be the letter Xi and its connotations leading to China and its President Xi Jinping. On the contrary, cultural traits seemed more important. Language here was used as a proxy to sociocultural characteristics usually related to nations and ethnicities: The Mediterranean countries (Spain, Portugal, Greece, and Romania) show a strong preference for pictorial images, as opposed to the North European (Germany, the Netherlands, Poland, and Sweden) which favor textual images. Similarly, Southeast Asian cultures (Hindi, Tamil, Tagalog, Malaysian, Urdu) also appear more iconocentric, while Far Eastern cultures (Chinese, Japanese, Korean) favor textual images (Table 1.2).

A word of caution is necessary at this point: the query to Twitter API was made with the hashtag #omicron, and therefore no country- and language-specific hashtags were used. Had we the opportunity to apply them as well, a better and more reliable understanding of the ethno-cultural and ethno-linguistic diversity in preference for either 'textual' or 'pictorial' images would be achieved. But this task is asking for new research, perhaps based on data from a longer period.

Personal characteristics may also participate in shaping the choice between 'textual' and 'pictorial' images to accompany a tweet. Such characteristics which can be traced within the tweet itself are the sentiments, emotions, and feelings. Though emotions were found to be dynamic and changing according to the evolution of the pandemic in each country, as well as to the measures taken, in our case only a snapshot is made, which may be further analysed in comparison to previous snapshots or to longer periods during the pandemic.

Table 1.2 Percent of textual and pictorial images per language

Language	Textual images	Pictorial images
Arabian	37	63
Catalan	18	82
Chinese	61	39
Dutch; Flemish	49	51
English	30	70
Esthonian	26	74
Finnish	29	71
French	28	72
German	40	60
Greek	12	88
Haitian	13	87
Hindi	18	82
Italian	23	77
Japanese	61	39
Korean	99	1
Malaysian	13	87
Norwegian	2	98
Polish	59	41
Portuguese	12	88
Romanian	11	89
Russian	35	65
Spanish	14	86
Swedish	81	19
Tagalog	19	81
Tamil	16	84
Thai	61	39
Turkish	28	72

Comparing the emotions expressed in tweets using ‘textual’ or ‘pictorial’ images has shown that the latter were prone to fear and sadness while the former expressed relatively stronger trust (Fig. 1.8).

Exploring deeper to the feelings produced by combinations of emotions, according to Plutchik’s theory, the use of pictorial images is related with the expression significantly more anxiety, guilt, pessimism, submission, modesty, awe, aggression, despair, and shame (Fig. 1.9). Hence, it is evident that the use of ‘textual’ or ‘pictorial’ images may be related to the emotional state of the individual. The former were related to tweets ex-pressing trust and hope, while the latter with tweets expressing fear and anticipation, anxiety, pessimism, and hope along with despair.

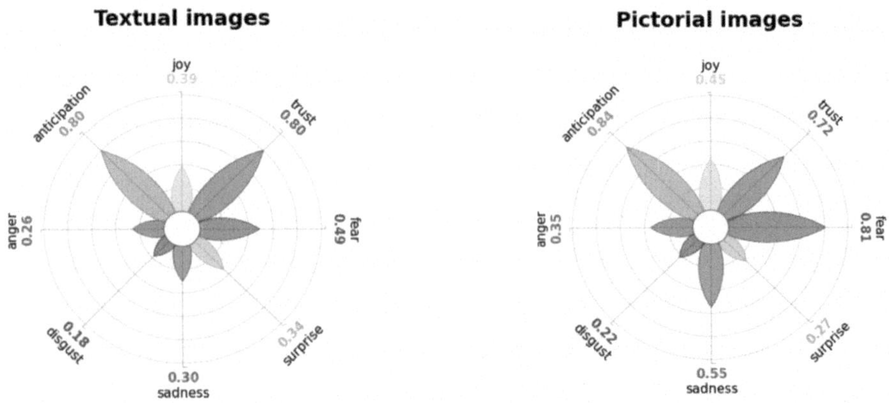

Fig. 1.8 Comparison of the emotions related to textual versus pictorial images

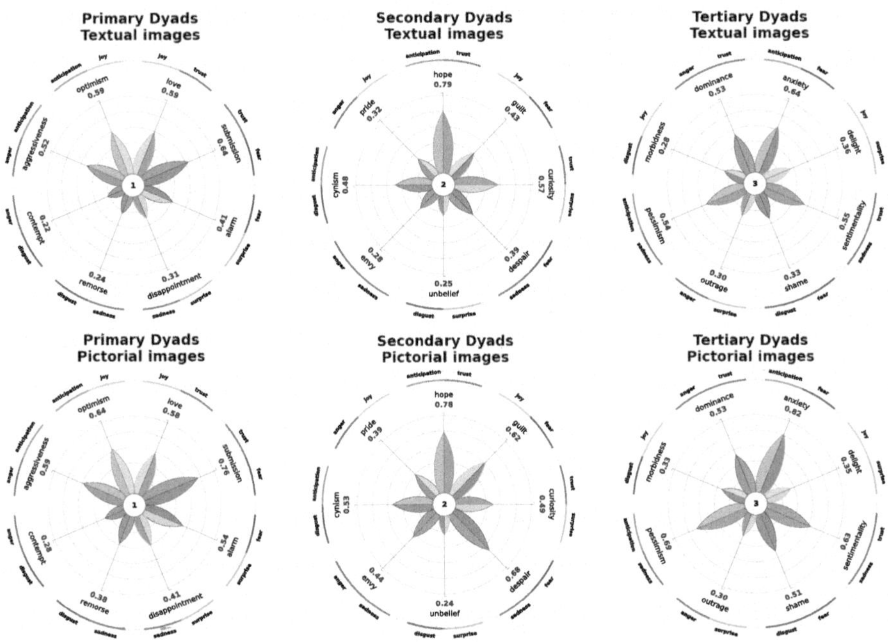

Fig. 1.9 Comparison of the feelings related to textual versus pictorial images

1.4.2 Categories of 'pictorial' Images

Delving deeper into 'pictorial' images was a necessary next step. The question guiding this step could be expressed as follows: "Do pictures with diverse formal characteristics reveal diverse emotional states of their users?" The four clusters produced by hierarchical clustering provided the ground for answering this question. We followed the same procedure with the previous step, in relating tweets with image clusters and computing the average emotions for each cluster. The result is illuminating since emotional differences between clusters are pronounced. The comics/cartoon cluster was related to more positive sentiment than any other cluster. The cluster with virus images expressed the less negative sentiment. On the other edge, infographics cluster was related to tweets with the most negative sentiment. Finally, the portrait cluster was balancing between positive and negative sentiment more efficiently.

Each group displayed a distinct emotional 'fingerprint.' Anticipation, fear, sadness, and trust were the most pronounced emotions in any cluster, though their importance was different for each one (Fig. 1.10).

The comics/cartoon cluster was related to tweets high in anticipation and low in anger, sadness, surprise, and even fear. The tweets expressed ironic joy or happiness ("Happy Omicron!"), or a playful relation to the *Futurama* story ("Protect yourself from the Omicron Persei 8 variant," or "I told my family there is a new variant called Omicron and my daughter said 'Lrrr'"). Thus, they were more optimistic, and underestimated fear and anxiety over the new virus variant.

The cluster with the virus images scored even lower in anger, anticipation, disgust, and surprise. On the other hand, the expression of fear was quite high. Thus, this cluster had a rather optimistic but still realistic perspective, without downgrading its dangers.

The infographics cluster scored high in anger, disgust, fear, and sadness. Negativity and negative sentiments take the form of fear and anger. Sometimes, the tweets related the name Omicron to occult traditions and sources ("Does anybody else hear the Omicron variant and immediately think of the Necronomicon in the Evil Dead?") in film, games like Destiny 2 characters and gear ("In case you needed a primer: Omicron: Omolon: #Omicron #omolon"), or anticipate danger ("Omicron Variant

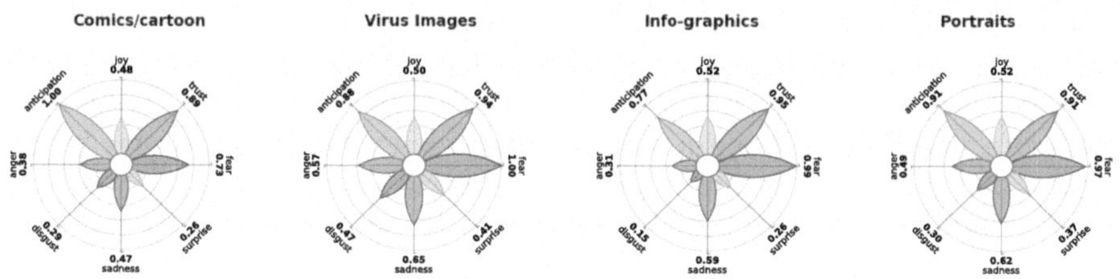

Fig. 1.10 The emotional fingerprint of the four pictorial clusters

May Be More Terrible Than Delta, more infectious and immune escape-capable," "'Omicron': How fear of new unstoppable Covid variant gripped the world").

The portrait cluster was characterised by the lack of extremities, either for good or for bad. As it was populated by almost two thirds of the corpus, it is possible that the images are related to tweets subscribing to diverging strategies. Therefore, it would be reasonable to delve deeper into its analysis, searching for sub-clusters and their emotional weights.

1.5 Discussion

Social media have been consistently used in political communication (e.g., during pre-election periods [2, 8, 10, 19] to 'listen' to the voice of the 'people' [15, 25, 44], or diffuse fake information aiming at influencing voting behaviour [18, 28, 43, 45]. In critical moments (e.g., during natural disasters) they are invaluable offering the authorities hints about needed interventions to cope with casualties, gauge popular morale, encounter false information etc. [30, 63]. During the pandemic social media were playing several roles: they offered to authorities and organizations of civil society information about issues that required their intervention (e.g., helpless people during lockdown), motifs of conforming or opposing to measures taken by the state (e.g., masks, social distancing, restrictions to movement), confronting the spread of misinformation and conspiracy theories, calibrating public policies and public discourse to manipulate emotions etc. As the crisis of the pandemic is lasting for more than 2 years, it may be used as a testbed for theories, as well as for experiments and explorations into social behaviour. So far, we presented the findings of a 'snapshot' in the history of reactions to official discourses regarding the pandemic. When the omicron variant of COVID-19 was announced, the pandemic was already in its eleventh month. People around the globe had been through periods of isolation, restrictions, promises for a return to their normal life, frustration, hopes for medicines or vaccines, and announcements of new variants. Thus, the announcement of a new variant had become almost a routine, though it was followed by renewed fear and uncertainty.

The analysis of collected Twitter data presented so far offered the opportunity to focus on the connection between such behavioural patterns and the underlying emotions and the visual content of social media. Previous research on the connection between multimodal (image or video and text) content in Twitter came mostly from the disciplines related to computer science, therefore their focus was on predicting rather than on describing social reality in its complexity. The choice made to follow the suggestion for a constant change of scale, the zooming in and out to the corpus, the transition from the macro- to the microscopic level, proved fruitful. We were able to detect the possibility of cultural background in the choice between 'textual' and 'pictorial' images. Cultures with a traditional place for imagery, where iconolatry won the battle during the late Middle Ages (i.e., the Greek-Orthodox and Roman-Catholic European South) have shown a preference for using 'pictorial' images, while iconoclastic cultures (i.e., the Protestant North) are inclined towards 'textual' ones. Of course, this is not an iron law either: Roman-Catholic Poland follows the Northern European paradigm, while the Protestant Norway leans enormously towards 'pictorial' images. What such canons and deviations reveal is that we are far from a holistic answer to the 'text' versus 'image' or 'rationalism' versus 'sentiment' dichotomies. There is not an answer suitable for all cases. Culture plays an important role, and even in our secular societies the imprint of religious past still holds.

The connection between image choice and emotions during a critical period proved fruitful as well. We have shown that within a single linguistic milieu visual rhetoric complements textual or verbal rhetoric and provides a rich repertoire to express emotions and feelings. It is true that English language tweets may originate in diverse states with significant socio-cultural differences, although more of them are based on Anglo-Saxon heritage. Therefore, more focused research might provide clues to internal differentiation within this corpus. In any case, computer vision techniques produced meaningful clusters of similar imagery, which were related with contrasting mix of emotions and feelings. Tweets using 'textual' images were more optimistic and expressed higher levels of trust and hope, whilst those preferring 'pictorial' images higher levels of fear and pessimism. The clusters within the 'pictorial' group of images revealed a joyful play with animated stories as an apotropaic mantra against the pandemic and its new variant; the choice of others to depict the 'enemy' virus expressing their decisiveness to confront it; the use of information related infographics and media clips as a means to support fears along with a vague feeling of being subject to hidden forces beyond their reach, the necessary sub-stratum of conspiracy theories; and finally, the turn to portraits of eminent or ordinary citizens transforming them into role-models, in an attempt to moderate uncertainty and fear, to anticipate for what would follow.

Such findings allow for a deeper understanding of the importance of visual content in social media during critical periods, be they natural disasters, pandemics, or political upheavals. The methodology followed is replicable in such situations and allows for a more delicate gauging of emotive dispositions, the flow of sentiments behind trivial expressions, the subtle subversion of textual meaning when contrasted with its visual counterpart. The constant whisper of ordinary people in social media, may produce either unintelligible noise, or distinct patterns guiding understanding of things untold.

1.6 Limitations

The methodology followed in the research described so far appears promising and generally applicable. On the other hand, two limitations have been already mentioned and require further attendance when applying either a similar research design or the tools and approach described above.

A first limitation lies in the conflation between culture and language. It was a workaround due to date shortcomings. Though Twitter provided information about country and location, most users neglect(ed) to disclose such information. Even when they did, they often answered vaguely (e.g., 'Somewhere'), according to their ideology (e.g., 'Planet Earth'), reading interests (e.g., 'Middle Earth'), or dual residence (e.g., 'Canada and Germany'). Language can be used as a proxy to nationality and national culture, especially in cases where nation and linguistic community are co-terminus. This is not the case with some of the most spoken languages like English, Spanish or French. As social media platforms are constantly changing the required data from users to open or maintain an account and the content available to researchers, the latter must pay attention to possibilities and limitations, and express concern about potential biases.

A second limitation is related to the way queries are conducted, and its impact upon data collection which may lead to biases. Though querying Twitter, or any other dataset, seems quite straight forward, it suffers from a hidden drawback: along with selecting data relevant to query terms, it leaves aside all data which don't match them. This is not far from taking what is irrelevant to the query terms as irrelevant to the issue under study. Understanding the difference between the two is crucial for alerting both the researchers and audiences for possible biases. In our research the query was conducted with the hashtag #omicron as the only search term, languages with non-latin alphabets may be under-represented in our corpus. Thus, the corpus may lean towards occidentalism and be West-centric. Another strategy would be to search for country- and language-specific hashtags to enrich the query terms. The resources needed for such a task (e.g. international research team, enhanced computational power) were not available. Thus, we confined ourselves to informing of such potential biases.

1.7 Conclusion

In this paper we wanted to demonstrate a methodology to identify the complex relationships between images and the psychological motivations of image-users, and even accomplish this task for large bodies of visual data. This methodology may be applied to different images: in the field of art one could to identify the emotions provoked by works of art to their audience, by mining the sentiment of published or unpublished texts about them (i.e. in monographs or posts in social media); in the field of science communication we could capture the emotions expressed by the public and the media whenever NASA publishes images of the universe, like the recent deepest Infrared Image of Universe yet, produced by the space telescope James Webb. Such a multi-level and collaborative research by different groups would allow us to study regularities and patterns in figurative and plastic features in images, which might be systematically related with certain sentiments, emotions, or feelings.

References

1. A. Adikari, R. Nawaratne, D. De Silva, S. Ranasinghe, O. Alahakoon, D. Alahakoon, Emotions of covid-19: content analysis of self-reported information using artificial intelligence. Journal of medical Internet research 23(4):e27,341 (2021)
2. R.H. Ali, G. Pinto, E. Lawrie, E.J. Linstead, A large-scale sentiment analysis of tweets pertaining to the 2020 us presidential election. Journal of big Data **9**(1), 79 (2022)
3. T. Arnold, L. Tilton, Distant viewing: analyzing large visual corpora. Digital Scholarship in the Humanities 34(Supplement_1):i3–i16 (2019)
4. F. Aslam, T.M. Awan, J.H. Syed, A. Kashif, M. Parveen, Sentiments and emotions evoked by news headlines of coronavirus disease (covid-19) outbreak. Humanities and Social Sciences Communications 7(1) (2020)
5. S.H. Awad, The social life of images. Vis. Stud. **35**(1), 28–39 (2020)
6. T. Bardakis, A semiotic approach to greek internet memes during the covid-19 pandemic. Punctum International Journal of Semiotics pp 31–43 (2021)
7. N. Bartley, A. Abeliuk, E. Ferrara, K. Lerman, Auditing algorithmic bias on twitter. In: Proceedings of the 13th ACM Web Science Conference 2021, pp 65–73 (2021)
8. M.K. Baxi, R. Sharma, V. Mago, Studying topic engagement and synergy among candidates for 2020 us elections. Soc. Netw. Anal. Min. **12**(1), 136 (2022)
9. M. Bernhard, Gugelmann galaxy: an unexpected journey through a collection of schweizer kleinmeister. International Journal for Digital Art History (2) (2016)
10. S. Boulianne, A.O. Larsson, Engagement with candidate posts on twitter, instagram, and facebook during the 2019 election. New Media & Society **25**(1), 119–140 (2023)
11. M. Broz, How many pictures are there: Statistics, trends, and forecasts. Photutorial (2024)
12. Y. Cai, H. Cai, X. Wan, Multi-modal sarcasm detection in twitter with hierarchical fusion model. In: Proceedings of the 57th annual meeting of the association for computational linguistics, pp 2506–2515 (2019)
13. J.F. Chartier, D. Pulizzotto, L. Chartrand, Meunier JG (2019) A data-driven computational semiotics: The semantic vector space of magritte's artworks. Semiotica **230**, 19–69 (2019)
14. G.S. Cheema, S. Hakimov, E. Müller-Budack, R. Ewerth, On the role of images for analyzing claims in social media. arXiv preprint arXiv:2103.09602 (2021)
15. P. Darius, Who polarizes twitter? ideological polarization, partisan groups and strategic networked campaigning on twitter during the 2017 and 2021 german federal elections' bundestagswahlen'. Soc. Netw. Anal. Min. **12**(1), 151 (2022)
16. J. Demšar, T. Curk, A. Erjavec, Č. Gorup, T. Hočevar, M. Milutinovič, M. Možina, M. Polajnar, M. Toplak, A. Starič, et al., Orange: data mining toolbox in python. the Journal of machine Learning research 14(1):2349–2353 (2013)
17. M.G. Dondero, Visual semiotics and automatic analysis of images from the cultural analytics lab: How can quantitative and qualitative analysis be combined? Semiotica **2019**(230), 121–142 (2019)
18. U. Dutta, R. Hanscom, J.S. Zhang, R. Han, T. Lehman, Q. Lv, S. Mishra, Analyzing twitter users' behavior before and after contact by the russia's internet research agency. Proceedings of the ACM on Human-Computer Interaction **5**(CSCW1), 1–24 (2021)
19. B. Enjolras, A. Salway, Homophily and polarization on political twitter during the 2017 norwegian election. Soc. Netw. Anal. Min. **13**(1), 10 (2022)
20. P. Favero, Learning to look beyond the frame: reflections on the changing meaning of images in the age of digital media practices. Vis. Stud. **29**(2), 166–179 (2014)
21. P. Hamilton, Representing the social: France and frenchness in post-war humanist photography. In: Representation: Cultural Representations and Signifying Practices, pp 75–150
22. R.T. Javed, M. Usama, W. Iqbal, J. Qadir, G. Tyson, I. Castro, K. Garimella, A deep dive into covid-19-related messages on whatsapp in pakistan. Soc. Netw. Anal. Min. **12**, 1–16 (2022)
23. M. Jay, Downcast eyes: The denigration of vision in twentieth-century French thought. Univ of California Press (1993)

24. M.T. Jerrentrup, Ugly on the internet: From# authenticity to# selflove. Vis. Stud. **36**(4–5), 491–506 (2021)
25. D. Karamouzas, I. Mademlis, I. Pitas, Public opinion monitoring through collective semantic analysis of tweets. Soc. Netw. Anal. Min. **12**(1), 91 (2022)
26. D. Kavanagh, The limits of visualization: Ocularcentrism and organization. In: The Routledge Companion to Visual Organization, Routledge, pp 64–76 (2014)
27. M. Klingemann, Rasterfairy. https://github.com/Quasimondo/RasterFairy (2015)
28. D.L. Linvill, P.L. Warren, A.E. Moore, Talking to trolls–how users respond to a coordinated information operation and why they're so supportive. Journal of Computer-Mediated Communication 27(1):zmab022 (2022)
29. L. Van der Maaten, G. Hinton, Visualizing data using t-sne. Journal of machine learning research 9(11) (2008)
30. S. Maleki, M. Mohammadalizadehkorde, A visual insight into tweeting activity before and during natural disasters: Case study of hurricane harvey. International Journal of Applied Geospatial Research (IJAGR) **12**(3), 1–15 (2021)
31. L. Manovich, Manifesto for democratic art history. Recuperado el 10 (2016)
32. L. Manovich, Cultural analytics. Mit Press (2020)
33. L. Manovich, A. Tifentale, M. Yazdani, J. Chow, The exceptional and the everyday: 144 hours in kiev. In: 2014 IEEE International Conference on Big Data (Big Data), IEEE, pp 72–79 (2014)
34. D. Marzana, C. Novara, N. De Piccoli, P. Cardinali, L. Migliorini, I. Di Napoli, E. Guidi, A. Fedi, C. Rollero, B. Agueli et al., Community dimensions and emotions in the era of covid-19. Journal of Community & Applied Social Psychology **32**(3), 358–373 (2022)
35. H. Metzler, B. Rimé, M. Pellert, T. Niederkrotenthaler, A. Di Natale, D. Garcia, Collective emotions during the covid-19 outbreak. Emotion **23**(3), 844 (2023)
36. S.M. Mohammad, P.D. Turney, Crowdsourcing a word-emotion association lexicon. Comput. Intell. **29**(3), 436–465 (2013)
37. F. Moretti, Conjectures on world literature. New Left Rev **2**(1), 54–68 (2000)
38. F. Moretti, The slaughterhouse of literature. MLQ: Modern Language Quarterly 61(1):207–227 (2000b)
39. C. Morin, A. Mercier, L. Atlani-Duault, Text-image relationships in tweets: Shaping the meanings of an epidemic. Societies **9**(1), 12 (2019)
40. C. Musello, Studying the home mode: An exploration of family photography and visual communication. Studies in Visual Communication **6**(1), 23–42 (1980)
41. C. Nau, J. Zhang, A. Quan-Haase, K. Mendes, Vernacular practices in digital feminist activism on twitter: Deconstructing affect and emotion in the# metoo movement. Fem. Media Stud. **23**(5), 2046–2062 (2023)
42. V. Negri, D. Scuratti, S. Agresti, D. Rooein, G. Scalia, A.R. Shankar, J.L.F. Marquez, M.J. Carman, B. Pernici, Image-based social sensing: combining ai and the crowd to mine policy-adherence indicators from twitter. In: 2021 IEEE/ACM 43rd International Conference on Software Engineering: Software Engineering in Society (ICSE-SEIS), IEEE, pp 92–101 (2021)
43. I. Neill Hoch, Russian internet research agency disinformation activities on tumblr: Identity, privacy, and ambivalence. Social Media+ Society 6(4):2056305120961,783 (2020)
44. A. Olteanu, A. Cernian, S.A. Gâgă, Leveraging machine learning and semi-structured information to identify political views from social media posts. Applied Sciences 12(24):12,962 (2022)
45. S. Park, S. Strover, J. Choi, M. Schnell, Mind games: A temporal sentiment analysis of the political messages of the internet research agency on facebook and twitter. New Media & Society **25**(3), 463–484 (2023)
46. L. Pauwels, A private visual practice going public? social functions and sociological research opportunities of web-based family photography. Vis. Stud. **23**(1), 34–49 (2008)
47. V.M. Pérez-Martínez, The message of pope francis in time of pandemic: Engagement and sentiment on twitter. Church, Communication and Culture **7**(1), 196–219 (2022)

48. R. Plutchik, H. Kellerman, Emotion, theory, research, and experience: theory, research and experience. Academic press (1980)
49. R. Plutchik, H. Kellerman, *The measurement of emotions,* vol. 4 (Academic Press, 1989)
50. D.M. Ponton, "never in my life have i heard such a load of absolute nonsense. wtf." political satire on the handling of the covid-19 crisis. Russian Journal of Linguistics 25(3):767–788 (2021)
51. S.L. Renshaw, S. Mai, E. Dubois, J. Sutton, C.T. Butts, Cutting through the noise: predictors of successful online message retransmission in the first 8 months of the covid-19 pandemic. Health security **19**(1), 31–43 (2021)
52. K. Simonyan, A. Zisserman, Very deep convolutional networks for large-scale image recognition. arXiv preprint arXiv:1409.1556 (2014)
53. Y. Skarpelos, Big visual data in social sciences. In: Computational Social Science in the Age of Big Data: Concepts, Methodologies, Tools, and Applications, Herbert von Halem Verlag, pp 235–265 (2018)
54. J. Sleigh, J. Amann, M. Schneider, E. Vayena, Qualitative analysis of visual risk communication on twitter during the covid-19 pandemic. BMC Public Health **21**, 1–12 (2021)
55. M. Stella, M.S. Vitevitch, F. Botta, Cognitive networks extract insights on covid-19 vaccines from english and italian popular tweets: Anticipation, logistics, conspiracy and loss of trust. Big Data and Cognitive Computing **6**(2), 52 (2022)
56. E. Summers, I. Brigadir, S. Hames, H. van Kemenade, P. Binkley, tinafigueroa, N. Ruest, Walmir, Chudnov D, Thiel D, Betsy, Chartier R, celeste, Lin H, Alice, Chosak A, Lenz M, McCain RM, Milligan I, Segerberg A, Shahrokhian D, Walsh M, Lausen L, Woodward N, eggplants, Ramaswami A, Nguyen B, Hereñú D, Milajevs D, Elwert F, Docnow/twarc: v2.14.0. https://doi.org/10.5281/zenodo.7799050, https://doi.org/10.5281/zenodo.7799050 (2023)
57. M.E. Unal, A. Kovashka, W.T. Chung, Y.R. Lin, Visual persuasion in covid-19 social media content: A multi-modal characterization. Companion Proceedings of the Web Conference **2022**, 694–704 (2022)
58. T. Underwood, A genealogy of distant reading. DHQ: Digital Humanities Quarterly 11(2) (2017)
59. A. Vempala, D. Preoţiuc-Pietro, Categorizing and inferring the relationship between the text and image of twitter posts. In: Proceedings of the 57th annual meeting of the Association for Computational Linguistics, pp 2830–2840 (2019)
60. X. Wang, J. Zheng, L.R. Tang, Y. Luo, Recommend or not? the influence of emotions on passengers' intention of airline recommendation during covid-19. Tour. Manage. **95**(104), 675 (2023)
61. B.F. Welles, On minorities and outliers: The case for making big data small. Big Data & Society 1(1):2053951714540,613 (2014)
62. F. Yao, X. Sun, H. Yu, W. Zhang, W. Liang, K. Fu, Mimicking the brain's cognition of sarcasm from multidisciplines for twitter sarcasm detection. IEEE Transactions on Neural Networks and Learning Systems (2021)
63. L. Zou, D. Liao, N.S. Lam, M.A. Meyer, N.G. Gharaibeh, H. Cai, B. Zhou, D. Li, Social media for emergency rescue: An analysis of rescue requests on twitter during hurricane harvey. International Journal of Disaster Risk Reduction **85**(103), 513 (2023)

Chapter 2
Digital Storytelling and First-Person Narratives in Heritage Interpretation: An Overview

Foteini Venieri

Abstract This chapter explores the evolving role of digital storytelling and first-person narratives in heritage interpretation. With roots in traditional storytelling and museum theatre, the practice has transformed through digital advancements such as Virtual and Augmented Reality, interactive platforms, and AI-driven narratives. These technologies enable more immersive, interactive, and personalized experiences, creating stronger emotional connections with audiences while addressing diverse perspectives. Focusing on the integration of first-person narratives, this chapter examines the potential of these methods to humanize history, foster empathy, and democratize heritage experiences. It also addresses the ethical and practical challenges associated with digital storytelling, including representation and the complexities of using AI. The discussion highlights opportunities for broader engagement and offers insights into best practices for leveraging digital tools in creating impactful heritage narratives.

Keywords Digital storytelling · First-person narratives · Heritage interpretation · Immersive experiences · Virtual Reality (VR) · Augmented Reality (AR) · AI · Museum theatre

2.1 Introduction

Over the last two decades, digital storytelling in heritage interpretation has gained significant attention [26, 28, 77, 82, 85–87, 102, 108]. Originating from the concept of 'museums as storytellers', storytelling in museums has been employed as a means to convey cultural values, historical events, and personal experiences, serving as a bridge between the past and the present [11, 41]. This approach has provided visitors with "context and sensory experiences in a way other places could not" [110].

F. Venieri (✉)
Hellenic Open University, Patras, Greece

Heterotopia, Athens, Greece
e-mail: f.venieri@heterotopia.gr

K. Karpouzis and Y. Skarpelos (eds.), *Computational Methods for the Digital Humanities*, Learning and Analytics in Intelligent Systems 57,
https://doi.org/10.1007/978-3-032-05099-1_2

However, the digital turn has significantly broadened the landscape of heritage storytelling, presenting new opportunities to reach an increasingly diversified audience [67, 78, 83, 94]. The scope and nature of storytelling in heritage interpretation have evolved significantly to adapt to this rapidly changing landscape.

Digital storytelling, just like traditional storytelling, invites the audience into a narrative space where they can actively engage with the stories being told, often through the eyes of historical figures or witnesses [6, 19, 109]. In physical settings, empirical data have shown that this approach has the potential not only to enrich the visitor experience but also to foster a multidimensional connection to heritage, allowing audiences to bridge their own experiences with complex historical narratives [46]. As the GLAM (Galleries, Libraries, Archives, and Museums) sector responds to an increasingly diverse and digitally connected audience, there is a growing emphasis on creating virtual experiences that cater to today's audience expectations: hyper-contextual information, interactive formats, and personalized content [110].

In this digital shift, museums are exploring ways to effectively transfer contextual and sensory experiences that traditionally take place in physical settings [22, 75]. Now that virtual environments increasingly either supplement or replace physical spaces, museums are developing multi -channel communication strategies that enable them to reach diverse visitor groups [25, 110]. Immersive digital storytelling (often coupled with multisensory experiences), direct engagement, and emotional connection, have become a core focus of heritage experience design [29, 76]. This blending of technological innovation with the art of narrative and performance–all the while drawing upon museological principles to ensure critical engagement and multiperspectivity- makes digital storytelling an increasingly interdisciplinary field. In this weaving together of different modes and media, practitioners must ensure the right balance and effective use of different professional and artistic languages.

A prominent form of digital storytelling that exemplifies these qualities is first-person interpretation. Originally rooted in museum theatre–a long-standing interpretation strategy in museums and heritage sites–first-person interpretation offers direct encounters with people from the past, and, more rarely, from the present [15, 46]. Actors or interpreters take on the role of a historical person, or a contemporary individual. They meet visitors and often engage in spontaneous dialogue, offering a sense of immediacy and intimacy which makes the past more relatable and vivid for the audience [46, 98, 105]. This mode of interpretation allows visitors to experience historical events and cultural practices from a personal perspective, offering direct encounters with people from the past, promoting empathy, multiperspectivity and a multi-level contextualization of the subject matter [46, 105]. Studies have also shown similar results in the digital environment: humanizing history and promoting relatedness and connections; thus bridging the gap between the present and the past [104, 104].

This chapter provides an overview of digital storytelling in heritage interpretation. It focuses on the role of first-person narratives, exploring both the historical background of storytelling in museums and the forms of first-person interpretation in digital storytelling today. In seeking to showcase the ways in which first-person narratives are employed to create more immersive and meaningful heritage experiences, this chapter will explore the opportunities and challenges they present. It will discuss the ethical considerations involved in digital storytelling and offer insights into future directions for this evolving practice.

2.2 Historical Background of Storytelling in Heritage Interpretation

Storytelling has become increasingly recognized as a central approach in heritage interpretation. Contemporary museums and heritage sites are now considered spaces where diverse stories, identities and perspectives can be discovered and shared [9, 11, 52]. Traditionally, these institutions were primarily repositories of artifacts, focused on preservation and presentation [1, 71]. However, in the mid-20th century, and especially with the advent of 'new museology' that highlighted museum's social role, they began evolving into interactive platforms for dialogue, inclusion, engagement, and education [40, 50, 103]. This shift has redefined contemporary museums as spaces where multiple perspectives and narratives intersect, promoting democratic values and encouraging critical engagement with heritage [74].

Heritage education and interpretation gained particular prominence after World War II and the establishment of the International Council of Museums (ICOM), which emphasized the role of museums in fostering cultural understanding and appreciation [38]. The focus shifted from static displays of objects to dynamic, visitor-centered interpretation, prioritizing the contextualization of information to reveal broader meanings and relationships within heritage. Freeman Tilden's seminal work in the 1950s further solidified this approach, describing interpretation as "an art that combines many arts" [100]. Tilden argued that effective interpretation requires creativity and narrative skills to make heritage accessible and meaningful, highlighting the importance of storytelling in engaging audiences.

Before the advent of digital technology, storytelling in heritage interpretation was closely tied to the spatial and temporal dimensions of the museum or site. The physical setting provided the environment in which stories were told, guiding the visitor's experience through carefully curated exhibitions and displays. This evolving approach in museums has emphasized creating a tangible connection between objects, narratives, and visitors, aligning with the paradigm shift towards more interactive and visitor-centered practices [3].

However, the digital era has introduced a new paradigm in heritage interpretation, fundamentally transforming how narratives are conveyed. Digital tools and methodologies allow heritage institutions to create more immersive and interactive experiences that go beyond physical spaces. Through digital media, museums can develop multi-layered narratives that engage visitors in virtual environments, enabling dynamic and personalized interactions with heritage. These technological advancements offer unique opportunities for museums to reach broader audiences and explore new forms of storytelling. However, they also raise challenges related to representation, factual accuracy, and the integrity of the stories being told, as well as numerous ethical considerations [18].

In this evolving landscape, museums and heritage sites are reimagining their roles, embracing digital storytelling as a means of expanding their interpretive practices and reaching a broader audience. By integrating digital media, these institutions can offer multi-layered narratives that transcend the limitations of physical space, creating virtual environments where stories can unfold in new and transformative ways [8].

2.3 Digital Storytelling and Technological Advancements in Heritage

Digital storytelling has redefined the ways heritage narratives are constructed and conveyed [80]. The digitization of collections, the creation of open-access resources, such as Europeana, and the rise of interactive digital platforms have given birth to new forms of storytelling, allowing for dynamic and immersive experiences that transcend traditional museum environments [69, 95]. In heritage contexts, digital storytelling uses a range of media, including Virtual and Augmented Reality (AR/VR), interactive websites, mobile applications, live online events, virtual tours, games, and interactive documentaries. These tools create interactive, often multi-sensory experiences that make heritage more accessible and engaging [21, 92, 101]. Digital storytelling can vary in structure; some narratives are linear, like cinematic experiences or VR films, while others embrace nonlinear, multi-layered storytelling that allows audiences to explore content in a personalized way [51, 109].

Artificial Intelligence (AI) is playing an increasingly important role in heritage storytelling. AI technologies allow institutions to create personalized experiences for users, tailoring content to match individual interests [36]. For instance, AI-powered virtual assistants and chatbots can provide real-time answers to visitor questions, mimicking the experience of live interpretation and offering a conversational way to engage with heritage content [35]. These innovations help bring digital storytelling closer to traditional in-person experiences by simulating interactions with guides or experts, bridging the gap between physical and virtual engagement [56, 58].

The COVID-19 pandemic further accelerated the adoption of digital storytelling in heritage, as museums and cultural institutions sought new ways to connect with audiences during closures. Virtual workshops, live-streamed events, and interactive online sessions became essential elements of the museum experience, enabling institutions to reach global audiences and foster a sense of continuity despite physical constraints [5, 104]. This shift also prompted heritage institutions to consider how digital engagement can capture the physicality and social aspects of in-person visits, essential components of traditional heritage experiences.

The museum sector has come up with various solutions in order to address these challenges. For example, the use of loan boxes that provide multisensory materials can help preserve a sense of physical engagement, while digital platforms can facilitate social interaction, fostering a sense of community among participants [105]. The inclusion of first-person narratives and diverse voices in digital storytelling enhances the interpretive process by offering multiple perspectives and encouraging a more nuanced understanding of heritage. In this way, the narrative process can become more democratic and challenge dominant historical discourses, creating space for underrepresented or marginalized voices to be heard and challenge dominant narratives [43, 83, 96].

However, the shift to digital storytelling in heritage interpretation is not without its complexities. The use of digital media requires careful consideration of how narratives are constructed and presented, ensuring that the digital representation of heritage does not distort or oversimplify complex cultural stories [39]. There is also the need to navigate the ethical dimensions of storytelling, particularly regarding the ownership and factual accuracy of narratives, and especially regarding the mediation of AI [61]. Museums and heritage institutions must balance between creating engaging digital content and respecting the cultural significance of the stories being told [18]. First-person narratives, in particular, demand a sensitive and ethical approach, ensuring that they are presented with respect for the individuals and communities involved [23].

Nevertheless, digital storytelling is positioned as a central element in the future of heritage interpretation. Initiatives such as the "Virtual Museum Mediation Toolkit" [45], Museum4punkt0's "Suggestions and Tools for Digital Cultural Mediation" [64], and "Remote Learning in Museums, Heritage, and Cultural Settings" demonstrate the potential of digital strategies in creating flexible, visitor-centered interpretation, promoting active participation and engagement. This shift toward more dynamic and inclusive digital storytelling opens up new opportunities for exploring historic characters in first-person within digital heritage. As digital narratives evolve to embrace a wider diversity of voices, first-person storytelling offers a way to craft immersive, personal experiences that deeply resonate with audiences, making history feel closer and more relatable.

2.4 Role and Impact of First-Person Narratives

First-person narratives in heritage interpretation are a subset of museum theatre, an interpretive strategy that originated at Skansen, the world's first open-air living history museum, established in Sweden in the late 19th century. Arthur Hazelius, the founder of Skansen, believed that objects lose much of their interpretive power when separated from their original context. To counteract this, he introduced costumed interpreters who reenacted daily life, creating an immersive historical environment that brought stories and traditions to life. This practice spread across Europe and eventually reached North America, where figures like George Francis Dow and Henry Ford used live interpretation in open-air museums to animate history for visitors [4]. By the mid-20th century, Freeman Tilden's work with the National Park Service formalized the principles of "living history," emphasizing the need for interpretive practices that reveal deeper meanings and connections beyond mere facts [100].

The social movements of the 1960's initiated a shift toward more inclusive narratives, through the development of social history and new museology, prompting museums to incorporate diverse voices and address complex social issues [55, 97]. This change also gave rise to more participatory forms of museum theatre, where audiences were encouraged to engage directly with historical figures, thereby facilitating dialogues on history, culture, and identity [2]. The founding of the International Museum Theatre Alliance (IMTAL) in the 1990's marked a significant step in formalizing museum theatre as a distinct practice, endorsing the use of drama and first-person narratives as methods to foster emotional and cognitive engagement [105].

Today, first-person interpretation is a core strategy in many museums, historic homes, and heritage sites worldwide, aiming to animate collections, spaces, and objects by revealing the intricate stories they represent [46, 59, 60]. These performances are often carried out by actors or interpreters and can be tailored for specific visitor groups or the general public. They encompass various formats, including monologues, interactive events, promenade performances, and site-specific enactments, each ranging from scripted to improvised forms involving solo or ensemble casts. They are inherently multimodal, engaging visitors through sight, sound, touch, and movement, occasionally incorporating reading and writing elements to deepen the experience and adapt to different learning styles.

Research into visitor experiences consistently indicates that museum theatre enhances enjoyment and learning [57], improves the overall museum experience [10, 65, 91], and aids in the retention and understanding of interpretive messages [63, 93]. Studies highlight that it encourages deeper involvement with subject matter [63], facilitates engagement with complex or abstract ideas [8], and evokes empathy and a sense of "time travel" [10, 63]. Furthermore, it helps visitors draw connections between exhibits and their own experiences, as well as to contemporary issues, making historical subjects more relatable and meaningful. This approach promotes "focused-looking" at exhibits [47], encourages historical thinking [48], and provides opportunities for interactive engagement and collaborative meaning-making through audience participation [31].

A three-year study by the University of Manchester [46] highlighted the capacity of museum theatre to give voice to marginalized communities, add elements of surprise and enjoyment, and effectively engage visitors with difficult topics, leaving a lasting impact. The study underscored that the quality of engagement depends not only on the performance itself but also on how the experience is framed, including pre- and post-performance activities. It also identified four key functions of museum theatre: illustrative (demonstrating skills or practices), explanatory (providing context), revelatory (highlighting lesser-known stories), and provocative (challenging assumptions and sparking debate). Museum theatre, especially through first-person interpretation, serves an essential function in conveying the "intangible human remains" of past societies–such as behaviors, attitudes, and social norms–within their specific historical, social, and economic contexts [34]. By shifting the focus from an object's intrinsic value to its broader cultural significance, museum theatre offers a multi-level contextualization that draws on intellectual, emotional, social, political, spiritual, and aesthetic dimensions. In this way it reframes objects and addresses the challenges of decontextualization by situating them within a framework of contrasting values and narratives [46].

Today, various museums across Europe and North America implement first-person interpretation to create immersive museums experiences. For example, Skansen Open-Air Museum in Stockholm, Sweden, continues its tradition of employing costumed interpreters to reenact daily life from different historical periods, allowing visitors to interact with historical figures such as farmers, craftsmen, and shopkeepers. In the United Kingdom, Beamish Museum showcases life in Northern England during the early 20th century. Interpreters embody the roles of miners, shopkeepers, and other townspeople, providing an immersive introduction to the social and economic fabric of the time. The Archeon Museum Park in the Netherlands recreates different periods from Dutch history, including the Stone Age, Roman times, and the Middle Ages, where interpreters wear period-specific clothing and reenact various aspects of daily life. Visitors can engage directly with these historical figures, gaining insights into the skills, traditions, and lifestyles of past communities. The Tenement Museum in New York City brings the lives of immigrant families to life by having interpreters embody real historical residents, creating a direct link to the resilience, adaptation, and community of immigrant experiences [20]. At Colonial Williamsburg in Virginia, USA, interpreters immerse visitors in the American colonial period, portraying figures from all walks of life–artisans, enslaved individuals, political figures, and townsfolk–providing an engaging look at the complexities of 18th-century society and the roots of American independence.

2.5 The Digital Transition of First-Person Narratives

First-person interpretation in digital storytelling within heritage contexts manifests through various forms, including VR and AR experiences, interactive digital platforms (such as websites and mobile applications), on-line live programming, interactive exhibits, and videos on social media platforms like YouTube, TikTok, and Facebook. These narratives are often brought to life by actors, avatars, or digital humans: characters designed with realistic, human-like appearances [16]. The narratives can follow linear or non-linear paths, with content either pre-scripted by designers or dynamically generated by Artificial Intelligence. Some experiences are live, allowing real-time interaction, while others are asynchronous, offering flexible engagement. In some cases, geolocation technology is incorporated, particularly within mobile applications, to create interactive and playful environments, especially suited to urban settings [88].

All of these formats can support transmedia storytelling, which allows narratives to unfold across multiple platforms–VR experiences, mobile apps, social media, and interactive exhibits–creating a cohesive and layered "narrative universe." Through this approach, heritage sites can adapt content to various formats, meeting diverse learning styles and institutional objectives [8, 13]. At its best, transmedia storytelling aligns with contemporary museum goals by offering multi-layered contextualization, personalizing the visitor experience, fostering multiperspectivity by presenting varied historical viewpoints, and encouraging both emotional and critical engagement with the content (Figs. 2.1 and 2.2).

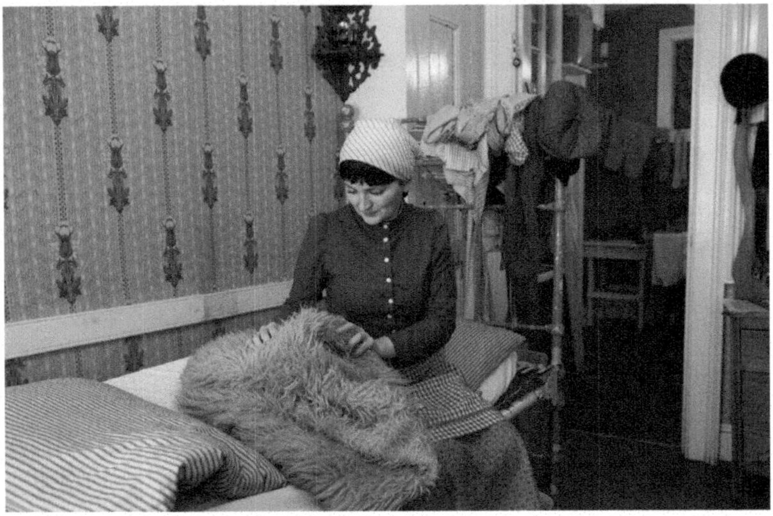

Fig. 2.1 Lower east side tenement museum©. Interpreter: Jessica underwood varma

Fig. 2.2 Slave trade at Colonial Williamsburg©. Interpreters (from left to right): Frank Megargee, Daniel Cross, Joseph Ziarko, Sharon Forsher, Jared Armstead), Jeremy Armstead, Kymberly Thomas, Christopher Allen, Suzy Allen, Brian Murray, Samuel Miller, and Willie Wright

2.5.1 Virtual Reality

Virtual Reality (VR) has become highly popular in the past decade for its ability to fully immerse visitors in digitally reconstructed settings, allowing them to explore historical environments and interact with characters and objects in real time. As Schofield et al. (2018) explain, VR encompasses two main approaches in cultural heritage: one uses real-time 3D graphics for fully interactive experiences, where users can move around and engage with objects independently; the other, known as cinematic VR (cVR), relies on high-quality, pre-rendered 360-degree video (either live-action or CGI), offering limited interactivity as users can only select the viewing angle rather than directly influence the narrative [27].

First-person interpretation in VR can establish a strong personal connection to historical figures. For instance, the "Anne Frank House VR" app transports users to the Secret Annex where Anne Frank and her family hid during the Nazi occupation. The app offers both a story mode and a tour mode, beginning with a brief introduction featuring photos and documents, before guiding users through the Annex. In the story mode, users hear Anne's voice describing her thoughts, with certain objects triggering diary entries projected onto the walls. Combined with soundscapes, movement, and visual cues within the reconstructed 3D environment, this approach fosters both immersion and a deep emotional connection to Anne's story, with Anne herself as the virtual guide. Notably, her presence is conveyed solely through her voice, with no avatar, which adds an intimate, audio tour quality to the experience (Fig. 2.3).

An even more immersive example is "Meeting Rembrandt," launched in 2017 by Force Field Studios in collaboration with the Rijksmuseum and Samsung Gear VR. This VR experience features a live actor playing Rembrandt, set against computer-generated environments that recreate his home and surroundings in 17th-century Amsterdam. Utilizing archival documents and artifacts, the production offers an authentic immersion into Rembrandt's life, enabling users to explore his house, interact with his studio, and even pose as models. Designed to be accessible without additional sensors or controllers, "Meeting Rembrandt" has received praise for its ability to make history tangible and relatable, and was awarded the Best Educational VR Award at the Cinequest Film and VR Festival in 2018 [62] (Fig. 2.4).

Fig. 2.3 Anne Frank house VR. *Source* https://www.annefrank.org/en/about-us/news-and-press/news/2019/7/4/renewed-vr-tour-anne-franks-secret-annex/

Fig. 2.4 "Meeting Rembrandt". *Source* https://vertigo-games.com/games/meeting-rembrandt/

The Louvre Museum's "Mona Lisa: Beyond the Glass," launched in 2019, similarly uses VR to provide a personalized experience that takes visitors back to the Renaissance. The VR journey, accessible on smartphones, lets users meet Mona Lisa–historically, Lisa Gherardini–who explains her attire, hairstyle, and the process behind her famous portrait. The experience was developed as part of the Museum's Leonardo da Vinci exhibition, aimed at offering a crowd-free, intimate engagement with the painting (Fig. 2.5).

2.5.2 Augmented Reality

Augmented Reality (AR) enhances real-world settings by overlaying digital elements onto physical environments, adding layers of historical narratives. The England's Historic Cities app uses AR across 12 historical sites, featuring characters like Shakespeare as guides. Although current AR applications in this setting tend to focus on well-known figures, they highlight the potential for incorporating a wider range of voices, including marginalized perspectives [24] (Fig. 2.6).

In 2015, at the Hart Cluett Mansion in New York, "an AR project allows visitors to play the role of a new servant in the Hart-Cluett House during the 1850's, working on a first day of trial employment. In order to secure a permanent position in the household, the new servant must execute a series of tasks assigned to them by the Harts' cook. These tasks include the delivery and retrieval of a variety of historic objects from various locations in the house, such as delivering a cup of tea to a member of the family, and fetching a chamber pot for emptying and cleaning."

2.5.3 Interactive Digital Platforms and Mobile Applications

Interactive digital platforms and mobile applications also serve as versatile tools for delivering first-person narratives. An example of such a mobile application, "RevAthens", recreates aspects of Athenian life during the Greek Revolution of 1821,

Fig. 2.5 "Mona Lisa: Beyond the Glass". *Source* https://www.louvre.fr/en/explore/life-at-the-museum/mona-lisa-beyond-the-glass-the-louvre-s-first-virtual-reality-experience

Fig. 2.6 Screenshot from the England's historic cities App trailer

Fig. 2.7 Screenshot from the REVAthens App

allowing users to meet historical characters in the form of avatars. Through digital storytelling techniques, users explore revolutionary Athens from the perspectives of real historical figures, enhancing the learning experience by highlighting marginalized voices and alternative narratives [37] (Fig. 2.7).

Interactive documentaries, or i-docs, also engage audiences through digital interactive technologies, allowing them to control the flow of information and narrative. Unlike traditional documentaries, i-docs are non-linear and integrate diverse media, such as video, sound, and data visualizations, providing layered content to enhance the user experience. This format enables users to become co-creators of their journey, deciding how to explore the content. For instance, the i-doc "Despo" was developed as a digital museum theatre experience. Named after its main character, Despo, is set in the 1960's in Soufli, Greece. Despo is a fictionalized yet historically grounded character, created through extensive research into the intangible cultural heritage of Thrace, including music, dance, local dialects, and traditions. The research was provided by the Friends of Music Society. The documentary allows users to engage with Despo as she shares stories about her life and surroundings. Visitors are invited to interact with her through a virtual interface simulating her courtyard. Through this medium, they can explore themes such as traditional women's songs and gendered cultural practices, making choices that shape the narrative pathway (Fig. 2.8).

This interactive approach, blending theatre, audiovisual media, and digital technology, aims to create an immersive experience where users actively participate in the storytelling. The project's design allows viewers to pause, choose specific stories, and explore additional cultural insights through supplementary materials, like definitions of local terms. This format offers users control over the narrative experience, thus enhancing accessibility and relevance [106].

Fig. 2.8 Screenshot from the interactive documentary "Despo". Interpreter: Katerina Holi

2.5.4 Digital Avatars, AI-Driven Narratives and Interactive Installations

Digital avatars and AI-driven narratives provide dynamic, interactive experiences by allowing historical characters to respond in real time. The USC Shoah Foundation's "Dimensions in Testimony" project, a collaboration between the USC Institute for Creative Technologies, the USC Shoah Foundation, and Conscience Display, is an example of this practice that combines AI-generated conversations with Holocaust survivors' avatars [7]. Visitors can ask questions that prompt real-time responses from pre-recorded videos that are drawn from the Foundations' oral history archives. According to the project's website, this "pioneering project integrates advanced filming techniques, specialized display technologies and next generation natural language processing to create an interactive biography" (Fig. 2.9).

Interactive installations in museums use digital tools such as touchscreens, motion sensors, and digital projections to allow hands-on interaction with first-person narra-

Fig. 2.9 Screenshot from USC Shoah foundation's "Dimensions in Testimony" project–conversation with Pinchas Gutter

tives. For example, the Canadian Museum for Human Rights features exhibits where visitors explore human rights stories through the voices of individuals who lived them, creating a strong personal connection to complex topics [12].

2.5.5 *Historical Chatbots, Social Media and Messaging Platforms*

Historical chatbots and social media platforms are increasingly being used to share first-person narratives with broader audiences. For example, the Anne Frank House employs a chatbot powered by deep learning AI (msg.ai) to provide personalized guidance and historical insights through instant messaging. Importantly, this chatbot does not impersonate Anne Frank but instead serves as a sort of digital docent.

While AI-powered applications that enable users to "speak" with historical figures are popular online, they have not gained significant traction, yet, in the GLAM (Galleries, Libraries, Archives, and Museums) sector, probably due to the ethical and practical challenges they present. Additionally, chatbots in these contexts typically rely on text-based communication and do not incorporate visual animations or performances.

TikTok, on the other hand, offers short videos and has become a popular platform for engaging younger users with historical content in creative and accessible ways. Two notable examples are the National Archives of the UK and the Black Country Living Museum, also in UK, both of which use TikTok to bring history to life through short, engaging videos that draw on first-person storytelling. Both initiatives have expanded the reach of traditional heritage storytelling, attracting millions of views and sparking dialogue among users. The National Archives of the UK leverages TikTok to share stories from Britain's historical records, often bringing historical figures to life in a relatable and humorous manner. These short, first-person videos allow viewers to experience historical events as if they were hearing directly from the people involved. For example, actors embody figures from different eras, recounting notable events or sharing personal stories from their historical context. This approach has included figures from the Tudor period and WWII, with actors dressing in period costumes and speaking as though they are the characters themselves. By presenting these stories through TikTok, the National Archives captures the attention of younger audiences, who are drawn into history in a personal and relatable way, making it feel vivid and immediate (Fig. 2.10).

The Black Country Living Museum has also gained significant popularity on TikTok, with a total of 1.2 million followers in November 2024, by creating first-person videos that recreate daily life in Britain's industrial past. Interpreters, dressed in period costumes, portray historical characters from the Black Country's history, including miners, factory workers, and shopkeepers, directly addressing the audience as though they were living in the 19th or early 20th centuries. These videos often feature humorous skits and use popular trends to engage viewers, all while convey-

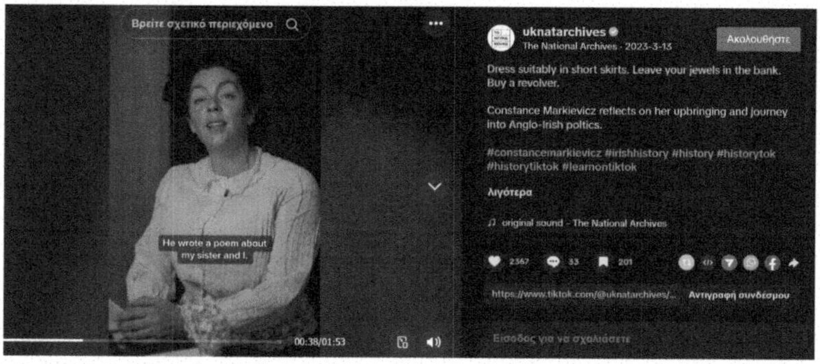

Fig. 2.10 Screenshot from the national archives UK Tik Tok channel. A costumed interpreter brings to life Constance Markievicz. Interpreter: Kathy Hipperson

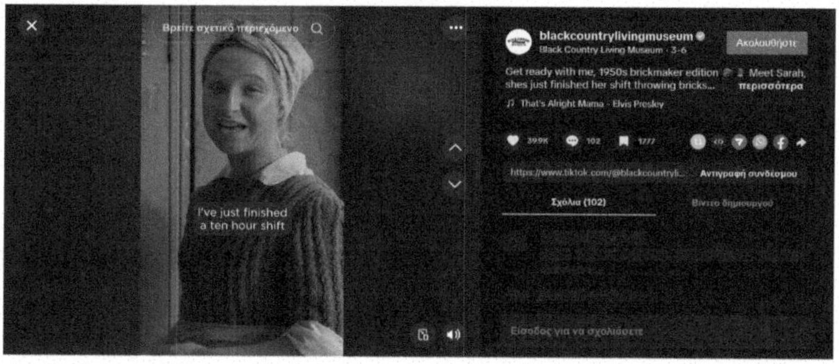

Fig. 2.11 Screenshot from the black country living museum TikTok channel. A costumed interpreter brings to life Sarah, a worker at the Cricket Field Brickworks. Interpreter: Amy Young

ing educational content. For example, interpreters discuss life during the Industrial Revolution, providing insights into working-class life, social customs, and regional dialects. The museum's success on TikTok has demonstrated how first-person storytelling on social media can turn historical content into a form of "edutainment," where audiences not only learn about the past but also connect with it emotionally and culturally (Fig. 2.11).

2.5.6 *Online Live Museum Programmes*

In the context of online live programming in museums and archives, museum theatre serves as an interactive tool for historical interpretation, where actors portray historical figures to guide virtual tours, perform scripted scenes, or present interpretive content. Like traditional museum theatre, the digital version enables dialogical direct encounters with people from the past. These performances also include multimedia elements, aiming to foster engagement and support critical thinking by offering diverse perspectives [104].

For example, the online live museum theatre programme "In the Footsteps of Vangelis Soukas" at the Museum of Greek Folk Musical Instruments demonstrates the power of digital storytelling to reach and engage marginalized communities. This programme was designed to address a lack of visibility and representation of Roma histories in the Greek cultural space. The programme uses a first-person narrative to introduce students to the life of Vangelis Soukas, a celebrated Roma musician, by allowing them to interact with Kostis, a fictional character and close friend of Soukas. Through Kostis's storytelling, students learn about Soukas's contributions to Greek music and the challenges he overcame, all within the context of their cultural heritage. This online, live narrative framework, made possible through the Museotek live learning and remote visits platform, allows Roma students to experience their community's history in a setting often inaccessible to them, fostering cultural pride and challenging stereotypes through personal and collective heritage [79]. The programme received a best practice award in 2024 from the Committee for Education and Cultural Action of the International Council of Museums (Fig. 2.12).

Another example of online live museum theatre focusing on archives this time is the "I Am Pilar" project. Developed as a cross-sector collaboration between Heterotopia, a not-for-profit company specializing in museum theatre, the Cultural Foundation of the National Bank of Greece, and the Museotek platform, the programme immersed students in post-war Athens, allowing them to engage directly with a fictional reporter, Sofia, preparing to interview the acclaimed Greek actress, Katina Paxinou. Through the Museotek platform, students participated remotely in a historically situated narrative that tasked them with supporting Sofia's research using English-language primary sources such as letters and press clippings from Paxinou's era. This design encouraged active learning, connecting students emotionally to historical content, while also engaging them cognitively through analysis and synthesis

Fig. 2.12 The costumed interpreter in the role of Kostis interacting with the students from the museum space. Source: Museum of Greek Folk Musical Instruments, Centre for Ethnomusicology 'Fivos Anoyanakis' Interpreter: Lefteris Katachanas

of archival materials. Notable examples of similar programmes also include the Tenement Museum in New York, the Jorvik Viking Centre in York, UK, and the Ceridwen Theatre Company's "Zoom Through History" in Wales [104].

2.6 Research Results

Research data regarding the impact of the above-mentioned projects are limited to the "Meeting Rembrandt" VR experience, the "RevAthens" app, the "Dimensions in Testimony" project, the Canadian Museum of Human Rights digital exhibits and the online live museum theatre programmes. However, broader research into digital storytelling based on first-person narratives provides some useful insights that shed further light on their impact and the issues they raise (Fig. 2.13).

Regarding the "Meeting Rembrandt" VR experience, users highlighted the realistic atmosphere and the personal connection fostered by this carefully crafted environment. They also emphasized the emotional impact of such experiences. For many, the experience evoked a sense of time travel and close personal connection to the artist, while it was praised for making complex subjects accessible and engaging [62]. These results are similar to research data from the evaluation of museum theatre in physical spaces [46].

Fig. 2.13 The costumed interpreter in the role of Sophia at the museum space. Another interpreter, serving as the host, facilitates the interaction and films Sophia. Source: Heterotopia. Interpreters: Katerina Holi, Nikoletta Dimopoulou

Evaluation results from the "REVAthens" app revealed a high appreciation of the app's dialogues and dramaturgy that brought to the surface aspects of the everyday life of the era that are not included in the usual, official narratives. Respondents with prior knowledge of the Greek Revolution found new insights, especially concerning the microhistorical aspects of daily life during the siege, which are often overlooked. Most participants indicated that the app provided a deeper understanding of historical events, particularly around interpersonal dynamics and the cultural diversity of the time. The app included elements of multilingual storytelling (Greek, Turkish, and Albanian), which reinforced a holistic view of Greek identity and social relations. The study found that participants valued the diverse representation of historical experiences, yet some felt the need for a deeper exploration of women's roles and other marginalized perspectives during the Revolution. On the other hand, users pointed to technical and navigational issues, already acknowledged by the research team that designed the app. These issues could not be met due to the limited funding available, which is likely to be an issue with other new media research and innovation projects [107].

The evaluation of the Shoah Foundation's project "Dimensions in Testimony" focuses on the usability and emotional engagement of an interactive storytelling system that shares Holocaust survivors' narratives. Early evaluations assessed the system's coherence and answer relevance using recorded survivor responses, with 95% of questions matched to responses. However, issues with speech recognition, data classification, and dialogue management led to occasional mismatches. Despite this, two-thirds of responses were accurate, while some "incorrect" answers still offered thematically relevant stories. Field testing was conducted with school groups at the Illinois Holocaust Museum, with survey questioning mediated by gallery docents. Observations showed positive audience engagement, with students demonstrating emotional connection and raising their hands for more questions. As students engaged further, they asked more in-depth questions, suggesting deepening interest. A comparative evaluation with live survivors revealed mixed responses: students valued the live interaction but appreciated the system's storytelling ability. Surveys showed a high sense of connection, with 72% feeling connected to the system's stories, though attention retention was slightly lower than with live survivors. Overall, both experiences had a positive impact on the students, with the interactive system enhancing educational and emotional engagement (Traum et al, 2015).

Regarding the digital exhibits at the Canadian Museum of Human Rights, Bidzinski et al [12] highlight how the Museum's oral history programme integrates audiovisual testimonies, allowing visitors to hear firsthand accounts of historical and contemporary rights issues, thus personalizing abstract concepts. Additionally, Failler [33] explains that interactive screens across the galleries, such as those in the Expressions Gallery, employ personal stories to evoke empathy and understanding among viewers. This approach aligns with the Museum's broader educational mission, using personal narratives to help audiences reflect on complex human rights topics [54]. Lavrence [53] also describes how digital storytelling techniques amplify emotional resonance, enabling visitors to engage directly with human experiences rather than viewing history from a detached perspective.

Research results for the live online museum theatre programmes have shown that the integration of first-person interpretation fostered high levels of engagement and motivation among students, as the performative element allowed them to "time travel" into historical settings, promoting empathy and a sense of personal connection to the past. Teachers reported that the use of live historical characters, fictionalized yet grounded in archival accuracy, helped students understand the complexities of historical events and figures. Students' interactions with the historic characters were not only entertaining but also educationally enriching, providing a space for dialogue and interpretation within a fictional yet historically rooted narrative framework [104]. The "In the Footsteps of Vangelis Soukas" programme was especially resonant with Roma students, fostering pride and a stronger sense of cultural identity. Teachers observed that many students related the material to their own lives, with the programme's dialogical approach deepening both understanding and engagement. By showcasing the contributions of Roma musicians to Greek heritage, the programme allowed students to explore history through a culturally relevant lens, increasing their interest in further historical exploration, including potential museum visits [79].

Apart from specific research results regarding the afore mentioned projects, there is a growing body of research that explores the impact and challenges of combining digital storytelling experiences with first-person narratives. A study at the National Museum of Korea evaluated three chatbot models: two "embodied" avatars (a contemporary docent and a historical character) and one messaging chatbot. The study found "that embodiment and reenactment strongly influence the three components of the museum experience". Specifically, embodiment alone contributes solely to knowledge/learning, while engagement and emotional connection are highly attributed to both embodiment and reenactment [66].

Previous research by Roussou [90] emphasized storytelling as a trigger of interest, empathy and imagination for visitors virtual cultural environments. The importance of "remarkable and unforgettable" human characters is highlighted as a necessary component for the development of an engaging narrative. In particular, according to Roussou, the challenge is to design the experience "formula" where characters tie into narratives that trigger users' imagination more than just visual sense, producing memorable experiences that successfully 'suspend disbelief' through narrative engagement.

Pietroni [75] underscores the importance of narration in bridging the past and present through museum objects. This is achieved by blending elements such as text, recitation style, music, sound effects, visual ambiance, camera movements, and rhythm. When these elements work together harmoniously, they not only enhance the emotional impact of the experience but also create a cohesive and engaging storytelling framework. She also notes that narration still represents the weakest link in the design of immersive experiences, since it rarely follows the precise rules from the original domains of cinema and theatre. Museum theatre directly addresses this gap by applying theatrical techniques within museum settings, skillfully integrating these narrative elements.

In the ArkaeVision project, that focused on the development of an "integrated system able to offer different modalities of exploitation of Cultural Heritage assets, including virtual representations of monuments, works of art and objects, as well as the stories associated to them" [14]—researchers applied a user-centric approach to create a more engaging user experience that includes elements of digital fiction and storytelling. In their project, the virtual scenarios and characters' features are designed to fuel users' immersion and emotional involvement, which in turn contributed to generating motivation and to activating the learning process more quickly. In relation to the same project, Pagano et al. [68] have shown how the integration of scenography, art and traditional film direction for representing the atmosphere of the past in a digital reproduction generated a strong emotional pull that impacted users' attention and their memory of the experience.

Discussing the role of storytelling in virtual museums, Sylaiou & Dafiotis (2020) draw attention to the importance of emotions as "fulcrums for engaged learning" and towards rethinking museums as affective spaces in dialogue with their audiences. Furthermore, research carried out within the EU-funded project EMOTIVE (Emotive virtual cultural experiences through personalized storytelling) explored how to design digital heritage experiences drawing on the power of "emotive storytelling" [73]. This research stressed the need to address emotions more explicitly in the design process, through drama-based affective narratives that are able to generate strong emotional engagement with the story, empathy with the characters, and connection with the sites. The main design indications for emotional cultural experiences identified by this project are: guiding visitors through a story in the present that makes them explore concepts of the past, creating stories that revolve around people and not exhibits, and supporting interaction between characters and real visitors [49, 73].

This approach is also exemplified in the project "Becoming Vincent" [17], which aimed to create an emotionally engaging virtual experience based on a narrative concept of Van Gogh's younger years. This project used the VR Oculus Rift system for creating an encounter to emotionally immerse the users into the painter's life, allowing them to take on the role of the artist himself and engage in a deep experience of the painter's heritage.

Research findings on the i-doc New Life, that narrates the life of refugees including oral testimonies as well, suggest that interactivity in digital storytelling enhances social engagement, with transmedia promotion, data-driven adjustments, and narrative quality identified as crucial for sustaining viewer interest [77].

According to Pujol-Tost [81] there is still a lack of a systematic evaluation of virtual heritage experiences making it difficult to assess the effectiveness of these narrative solutions beyond the usability/user experience level, despite the emphasis on users' emotional involvement.

2.7 Opportunities and Challenges

Digital storytelling holds great potential for increasing accessibility by breaking down geographic and social barriers, providing a platform for marginalized voices, and offering novel ways for visitors to connect with cultural narratives. This democratization of museum practices is especially significant in enabling underrepresented groups to shape and participate in cultural storytelling [72]. Additionaly, they allow audiences with disabilities or those who are geographically distanced to experience and interact with cultural heritage. Disabled users, for instance, have the opportunity to explore museum spaces remotely, as demonstrated by Evett and Tan [32], who found that immersive virtual environments allowed users to virtually "walk" through exhibitions and engage with artifacts otherwise inaccessible to them. Similarly, VR experiences are defined by their ability to physically immerse users and generate a psychological sense of presence within a virtual environment, factors shown to impact enjoyment, attitudes, and behavioral intentions [42]. Furthermore, the structure of these digital programmes incorporates feedback mechanisms through live interactions, or data collection, enabling significant insight into user interaction.

However, as digital storytelling brings new voices into cultural narratives, it also reveals complex issues of power dynamics and representation. The design and implementation of such projects must consider the influence of the creators on the narratives, as designers' perspectives and biases can inadvertently shape the portrayal of cultural heritage. Rouse [89] critiques the common oversight of failing to acknowledge "the power relations between designer and user," noting that this can result in digital storytelling projects that feel extractive rather than collaborative. To avoid these pitfalls, digital storytelling in heritage should follow participatory design principles that prioritize partnership, drawing on models from participatory heritage projects [96] and critical heritage studies. Such approaches emphasize the involvement of communities as co-creators in the storytelling process, fostering narratives that respect and more accurately represent the communities' lived experiences.

Helguera (cited in [89]) further underscores the importance of working closely with communities to prevent feelings of alienation or exploitation. Collaborative storytelling is achieved when communities are acknowledged as full partners, actively shaping the narrative rather than simply being depicted within it. Effective projects can incorporate continuous feedback mechanisms to allow represented communities to regularly assess and influence how their stories are told, ensuring cultural sensitivity and relevance.

Transparency and inclusivity are also key in digital storytelling. To avoid unintentional biases, designers should adopt transparent storytelling practices, clearly communicating the narrative intentions, limitations, and cultural assumptions inherent in their designs. Openly sharing these intentions helps audiences critically engage with the narrative and better understand the creator's perspective. Community-driven feedback systems further contribute to transparency by allowing represented communities to provide input and suggest adjustments to ensure ongoing cultural relevance and accuracy.

Another challenge in digital storytelling lies in achieving historical accuracy while maintaining audience engagement. Historical accuracy in representing cultural narratives is crucial for fostering an informed understanding of heritage, especially since digital storytelling techniques often reimagine past lives and events in ways that appeal to contemporary audiences. The International Council of Museums (ICOM) has provided ethical guidelines aimed at preserving authenticity and integrity in cultural representations, advocating for practices that avoid stereotyping or misrepresentation of marginalized groups [44]. Again, collaboration with community members and cultural experts ensures that narratives remain grounded in factual accuracy, thus fostering a culturally rich experience that honors the complexities of the past.

Emotional connection, emphasized in relevant literature for creating memorable and impactful visitor experiences, also needs to be framed in a way that allows relatability but also critical thinking. The term 'empathy paradox,' introduced by Jackson and Kidd [46], describes how empathy toward a character can sometimes overshadow the broader topic that the character represents, when emotional engagement is not carefully balanced with contextual interpretation. Along these lines, Raz [84] warns that while VR can deepen intersubjective affinity, it may not fully address the structural and historical contexts of marginalized experiences. Furthermore, VR's multisensory and embodiment effects particularly influence empathy, raising the risk of misleading and/or overestimating users' understanding of complex human experiences. Drawing from thinkers like Emmanuel Levinas, Raz also highlights the ethical necessity of recognizing the limits of empathy; the impression that a user "fully understands" may undermine the person's unique reality or even objectify him or her.

The integration of AI and virtual humans into digital storytelling introduces additional ethical considerations. Generative AI models, for instance, while they can analyze vast historical datasets and visitor feedback, as noted in the recent brief on Artificial Intelligence in the context of cultural heritage and museums by the European Parliamentary Research Service [70], can also simplify complex cultural narratives, risking oversimplified or homogenized representations of diverse cultural identities.

In a recent UNESCO report titled "AI and the Holocaust: rewriting history? The impact of Artificial Intelligence on understanding the Holocaust" [61], the following risks are addressed:

1. AI automated content may invent facts
2. Deepfake technology's potential to manipulate audio and video to fabricate content and spread hate speech. The report highlights that "deepfakes of celebrities have been used to spread Nazi ideology, or to simulate conversations with Nazi leaders including Adolf Hitler".
3. "Algorithmic bias" that "have led to some search engines and AI chatbots downplaying Holocaust facts or promoting far-right content, including Holocaust denial"
4. Oversimplification of history's complexity due to "AI's tendency to focus on the most well-known aspects of the Holocaust... The omission of lesser-known episodes and events in the history of the Holocaust reinforces stereotypical representations of the Holocaust and limits our understanding of a complex past which affected people in every country in Europe and in North Africa, and whose legacy continues to be felt worldwide." [61]

It also highlights that while "GenAI outputs may appear authoritative and accurate, AI systems lack human-level semantic understanding." Furthermore, AI systems can threaten privacy, as they can limit individuals' ability to control what information about their past is retrieved. AI systems also lack transparency, thus limiting our understanding of their functionality that would "hold AI companies accountable for possible errors in and harm from their systems. It is often unclear which data AI systems have been trained on or how algorithms decide what content to retrieve, generate, or prioritize." [61].

Though acknowledging the benefits that AI technologies bring to education and research, the report emphasizes that "it is essential for AI designers, policymakers, educators, and researchers to collaborate closely. Only AI systems equipped with robust safeguards and human rights assessments, coupled with an increased focus on developing digital literacy skills, can uphold the integrity of historical truth and ensure the responsible use of Artificial Intelligence"

The report also emphasizes the importance of digital media and AI literacy for students, educators, and heritage practitioners. Integration of ethical values, stakeholder consultation, and ongoing monitoring and evaluation of standards are considered vital aspects in combating the risks of AI.

Along the same lines, Taddeo and Floridi [99] propose an ethical framework for AI that emphasizes inclusivity, transparency, and accountability–principles that are essential for culturally sensitive storytelling. Museums that adopt AI-based storytelling methods must collaborate closely with cultural heritage experts to ensure that virtual representations respect the diversity and depth of cultural values. Additionally, AI systems should be developed with the capacity for transparency, where designers clearly communicate the limitations of AI narratives to users, helping audiences understand that while AI can enhance storytelling, it may not fully capture the subjective and nuanced nature of cultural identity.

The EU AI Act [30] further emphasizes the importance of limiting the risks associated with the use of AI, and provides a regulatory framework to uphold ethical standards, transparency, and accountability in the development and deployment of AI technologies within the heritage sector, ensuring responsible and equitable utilization of these powerful tools.

The adoption of digital storytelling in heritage contexts also requires substantial investment in technology, personnel, and training–resources that may not be readily available to smaller institutions. While digital platforms have been shown to significantly enhance visitor engagement and improve interpretive accuracy, the initial costs of VR, AI, and interactive mobile applications can be prohibitive. Nonetheless, collaborative models, such as the EU-funded EMOTIVE project, highlight how resource-constrained institutions can successfully implement digital storytelling through strategic partnerships and shared resources [73, 105]. Continuous investment in training and resource development is crucial to making digital storytelling accessible across various institutions.

2.8 Discussion and Conclusion

The evolution of digital storytelling in heritage interpretation, particularly through first-person interpretation, has reshaped not only how stories from the past are conveyed but also how they are experienced, understood, and valued in contemporary society. Rooted in museum theatre, first-person interpretation in digital contexts tries to balance the transition of techniques and methods, proven effective in the physical settings, to the digital realm. Either through physical or digital means, by seeing events through the perspectives of those who lived them, audiences can form a stronger, more personal connection with heritage. However, a critical question arises in this transition: Does this technological mediation transform historical experiences into more accessible cultural touchpoints, or does it risk commodifying heritage by translating it into consumable, potentially superficial interactions? While immersive technologies like VR, AR, and AI have expanded the reach and impact of first-person interpretation, their role in shaping collective memory and public history warrants closer examination. Can the dynamic, user-driven format of digital storytelling effectively balance personalization with historical accuracy and semantic content, or does it inevitably filter historical narratives through the subjective and sometimes biased lens of technological design? Are aesthetics and/or dramaturgical parameters and their impact examined thoroughly in the design and evaluation of such experiences?

Moreover, the ethical implications of integrating AI-driven narratives into first-person interpretation introduces significant complexities. Although AI enhances interactivity and personalization, does it adequately account for the cultural sensitivities and nuances that historians, museologists and cultural mediators bring to first-person interpretation? AI's reliance on training data raises concerns about perpetuating cultural biases, especially when working with narratives from marginalized or historically traumatized communities. Moving forward, it is crucial to explore how digital heritage projects can safeguard the integrity of these stories. What frameworks are necessary to guide the ethical implementation of AI in first-person interpretation and prevent the erasure or oversimplification of complex cultural identities? Should heritage institutions adopt standardized ethical protocols or develop flexible, project-specific guidelines to navigate the interplay of technology and cultural representation of historic characters and recently deceased people?

Finally, as digital first-person interpretation continues to grow, the field must also confront the question of sustainability. The substantial investments required for digital platforms, skilled personnel, and cutting-edge technology make digital forms of first-person interpretation potentially exclusionary, favoring well-resourced institutions. This raises a broader issue of equity within the heritage sector. How can smaller institutions and underfunded heritage sites participate in this digital evolution? What role could partnerships or open-source platforms play in leveling the playing field, ensuring that digital storytelling remains accessible to a wide spectrum of heritage institutions and communities?

Regardless of the significant impact digital first-person narratives have on visitor experience it is crucial to remain critically self-aware, adaptable, and responsive to the evolving relationship between technology, culture, aesthetics and historical interpretation. An ongoing dialogue between technologists, heritage professionals, artists and the communities these narratives represent, is crucial for ensuring that digital storytelling and first-person interpretation fulfills its promise as a tool for preserving and illuminating, rather than distorting or commercializing, our shared cultural heritage.

References

1. E. Alexander, *Museums in Motion: An Introduction to the History and Functions of Museums* (American Association for State and Local History, Nashville, 1979)
2. S. Alsford, D. Parry, Interpretive theatre: A role in museums? Museum Management and Curatorship **10**(1), 8–23 (1991)
3. G. Anderson (ed.), *Reinventing the Museum: Historical and Contemporary Perspectives on the Paradigm Shift* (AltaMira Press, Walnut Creek, CA, 2004)
4. J. Anderson, *Time Machines: The World of Living History* (The American Association for State and Local History, Nashville, 1984)
5. A. Antoniou, Mixed cultural visits or what covid-19 taught us. *Computers*, 12(7) (2023). Accessed May 6, 2024
6. G. Arrigoni, A. Galani, Recasting witnessing in museums: Digital interactive displays for dialogic remembering. Int. J. Herit. Stud. **27**(2), 250–264 (2020)
7. R. Artstein, D. Traum, O. Alexander, A. Jones, P. Debevec, Time-offset interaction with a holocaust survivor, in *Proceedings of the 19th International Conference on Intelligent Virtual Agents*. (Cham, Springer, 2016), pp. 6–13
8. N. Basaraba, *Remixing Transmedia for Cultural Heritage Sites: The Rhetoric, Creative Practice, and Evaluation of Digital Narratives*. PhD thesis (Trinity College Dublin, 2020)
9. P. Basu, Museum, landscape and the storytelling space between. Landscape and Arts **34**(35), 2–7 (2005)
10. L. Baum, C. Hughes, Ten years of evaluating science theatre at the museum of science, boston. Curator: Museum J. **44**(4), 355–369 (2001)
11. L. Bedford, Storytelling: The real work of museums. Curator: Museum J. **44**(1), 27–34 (2001)
12. H. Bidzinski, J. Giesbrecht, R.L. Hinther, S. Reilly, Building the oral history program at the canadian museum for human rights, in *Oral History Forum d'histoire orale*, volume 37 (2017)
13. M. Bourdaa, France: Telling tales of cultural heritage using transmedia storytelling, in *Global Convergence Cultures*. Taylor & Francis (2018)
14. G. Bozzeli, A. Raia, S. Ricciardi, M. De Nino, N. Barile, M. Perrella, M. Tramontano, A. Pagano, A. Palombini, An integrated vr/ar framework for user-centric interactive experience of cultural heritage: The arkaevision project. *Digital Applications in Archaeology and Cultural Heritage*, 15 (2019)
15. T. Bridal, *Exploring Museum Theatre* (Alta Mira Press, Walnut Creek, CA, 2004)
16. D. Burden, M. Savin-Baden, *Virtual Humans: Today and Tomorrow*. (Taylor & Francis, 2019)
17. L. Calvi, M. Hover, Becoming vincent: A multifaceted story in a multifaceted ecosystem, in *Museum Experience Design*. (Springer, 2018), pp. 279–300
18. F. Cameron, S. Kenderdine (eds.), *Theorizing Digital Cultural Heritage: A Critical Discourse* (MIT Press, Cambridge, MA, 2009)
19. V. Cesário, P. Campos, The integrated museum engagement model (imem): Bridging participatory design, immersive storytelling, and digital representation for enhanced museum experiences. The International Journal of the Inclusive Museum **17**, 63–81 (2024)
20. J. Chalfen, Performing human rights at two historic site museums, in *Performing Heritage: Research*. ed. by A. Jackson, J. Kidd (Practice and Innovation in Museum Theatre and Live Interpretation. Manchester University Press, Manchester, 2011)

21. E. Champion, Otherness of place: Game-based interaction and learning in virtual heritage projects. Int. J. Herit. Stud. **14**(3), 210–228 (2008)
22. L. Ciolfi, Embodiment and place experience in heritage technology design, in *The International Handbooks of Museum Studies* (2014), pp. 419–445
23. T. Copeland, *Constructing Pasts: Interpreting Heritage Sites* (Routledge, London, 2006)
24. J. Decker, A. Doherty, J. Geigel, G.D. Gacobs, Bridging past and present: Creating and deploying a historical character to engage audiences through ar and vr, in *RISE IMET 2021, volume 1432 of CCIS*. ed. by M. Shehade, T. Stylianou-Lambert (Springer, 2021), pp.139–155
25. C. Devine, M. Tarr, The digital layer in the museum experience, in *Museums and Digital Culture*. ed. by T. Giannini, J. Bowen. Springer Series on Cultural Computing. (Springer, Cham, 2019), pp.395–407
26. A. Di Russo, D. Papathanasiou-Zuhrt, The makhortykh and mann experiential digital heritage narrative in the black sea basin, in A. Kavoura, T. Borges-Tiago, and F. Tiago, editors, *Strategic Innovative Marketing and Tourism*, Springer Proceedings in Business and Economics. (Springer, Cham, 2024)
27. K. Dooley, Storytelling with virtual reality in 360-degrees: A new screen grammar. Studies in Australasian Cinema **11**(3), 161–171 (2017)
28. R. El Gamil, Storytelling as a tool for safeguarding and marketing the intangible cultural heritage (ich): The case of nubia city, egypt. Int. J. Tour. Res. **18**, 165–185 (2017)
29. K. El Raheb, L. Kougioumtzian, M. Stergiou, D. Petousi, A. Katifori, K. Servi, V. Kriezi, V. Vraka, S. Merakos, A. Charkiolakis, F. Venieri, M. Boile, Y. Ioannidis, Designing an augmented experience for a music archive: What does the audience need beyond the sense of hearing? Journal on Computing and Cultural Heritage **15**(4), 1–24 (2022)
30. European Parliament, EU AI Act: first regulation on artificial intelligence. European Parliament, Brussels (2023). https://www.europarl.europa.eu/topics/en/article/20230601STO93804/eu-ai-act-first-regulation-on-artificial-intelligence
31. S. Evans, Personal beliefs and national stories: Theatre in museums as a tool for exploring historical memory. Curator **56**(2), 189–197 (2013)
32. L. Evett, Y.K. Tan, Talk your way round-a speech interface to a virtual museum. Disabil. Rehabil. **24**, 607–612 (2002)
33. A. Failler, Canada 150: Exhibiting national memory at the canadian museum for human rights. Citizsh. Stud. **22**(4), 358–380 (2018)
34. A. Farthing, Authenticity and metaphor: Displaying intangible human remains in museum theatre, in *Performing Heritage: Research*. ed. by A. Jackson, J. Kidd. Practice and Innovation in Museum Theatre and Live Interpretation. (Manchester University Press, Manchester, 2011), pp.94–106
35. A. Galani, J. Kidd, Hybrid material encounters: Expanding the continuum of museum materialities in the wake of a pandemic. Museum and Society **18**(3), 298–301 (2020)
36. H. Gao, M. Lindquist, R. Serrano Vergel, Ai-driven avatars in immersive 3d environments for cultural heritage. *Journal of Digital Landscape Architecture* (2024), pp. 640–651
37. A. Gazi, Y. Stoyannides, F. Venieri, I. Marmaras, S. Ziff, Y. Scoulidas, T. Giannakis, Revathens: Bringing athens of the revolution to life through museum theatre methodology and digital gamification techniques, in *Emerging Technologies and the Digital Transformation of Museums and Heritage Sites*. ed. by M. Shehade, T. Stylianou-Lambert (Springer, 2021), pp.166–183
38. N. Gesché-Koning, Museum education and interpretation in the second half of the 20th century, in N. Gesché-Koning, editor, *History of Museum Education and Interpretation*. (ICOM Belgium Flanders, 2021)
39. E. Giaccardi, Collective storytelling and social creativity in the virtual museum: A case study. Des. Issues **22**(3), 29–41 (2006)
40. C. Gray, V. McCall, *The Role of Today's Museum* (Routledge, London, 2020)
41. E.H. Gurian, What is the object of this exercise? a meandering exploration of the many meanings of objects in museums. Daedalus **128**, 163–183 (1999)

42. D. Gutierrez, V. Sundstedt, F. Gomez, A. Chalmers, Modeling light scattering for virtual heritage. Journal of Computing and Cultural Heritage (JOCCH) **1**(2), 8 (2008)
43. C. Hughes, *Museum Theatre: Communicating with Visitors through Drama* (Heinemann, Portsmouth, NH, 1998)
44. International Council of Museums (ICOM). Code of ethics for museums, 2013. Accessed 5 Nov 2024
45. International Council of Museums (ICOM). Virtual museum mediation: A necessary adaptation of methods, 2022. Accessed 6 May 2024
46. A. Jackson, J. Kidd, Performance, learning and heritage: Research report. Centre for Applied Theatre Research, University of Manchester, 2008. Accessed 5 Nov 2024
47. A. Jackson, H. Rees Leahy, 'seeing it for real...?' authenticity, theatre and learning in museums. Res. Drama Edu. **10**(3), 303–325 (2005)
48. C. Jones, *An Illusion that Makes the Past Seem Real: The Potential of Living History for Developing the Consciousness of Young People*. PhD thesis (University of Leicester, 2011)
49. A. Katifori, M. Roussou, S. Perry, G. Drettakis, S. Vizcay, J. Philip, The emotive project - emotive virtual cultural experiences through personalized storytelling, in *Proceedings of CIRA@EuroMed* (2018), pp. 11–20
50. S. Knell, *The Contemporary Museum: Shaping Museums for the Global Now* (Routledge, London, 2019)
51. H. Koenitz, *Understanding Interactive Digital Narrative: Immersive Expressions for a Complex Time* (Routledge, London, 2023)
52. A. Langer, editor. *Storytelling in Museums*. Rowman & Littlefield Publishers / American Alliance of Museums, 2022
53. C. Lavrence, Emotion in the history of human rights: A case study of the canadian museum for human rights, in *The Routledge History of Human Rights*. ed. by R. Patel, K.P. Patel (Routledge, New York, 2019), pp.305–321
54. E. Lehrer, Thinking through the canadian museum for human rights. Am. Q. **67**(3), 885–892 (2015)
55. W. Leon, R. Rosenzweig, *History Museum in the United States: A Critical Assessment* (University of Illinois Press, Urbana and Chicago, 1989)
56. Y. Lian, J. Xie, The evolution of digital cultural heritage research: Identifying key trends, hotspots, and challenges. Sustainability **16**, 7125 (2024)
57. J.M. Litwak, A. Cutting, *Evaluation of interpretive programming* (Minnesota History Center, Saint Paul, MN, 1996)
58. O.M. Machidon, M. Duguleana, M. Carrozzino, Virtual humans in cultural heritage ict applications: A review. J. Cult. Herit. **32**, 249–260 (2018)
59. S. Maggelsen, *Living History Museums: Undoing History Through Performance* (The Scarecrow Press Inc., Maryland, 2007)
60. S. Maggelsen, R. Justice-Malloy (eds.), *Enacting History* (The University of Alabama Press, Alabama, 2011)
61. M. Makhortykh, H. Mann, AI and the holocaust: rewriting history? The impact of artificial intelligence on understanding the Holocaust. UNESCO, Paris (2024). https://www.unesco.org/en/articles/ai-and-holocaust-rewriting-history-impact-artificial-intelligence-understanding-holocaust
62. A. Marasco, Beyond virtual cultural tourism: History-living experiences with cinematic virtual reality. Tourism and Heritage Journal **2**, 1–16 (2020)
63. E.M. Munley, Evaluation study of 'buying freedom', in *Perspectives on Museum Theater*. ed. by C. Hughes (American Association of Museums, Washington, DC, 1993), pp.69–94
64. Museum4punkt0, Impulse & tools für die digitale kulturvermittlung [suggestions and tools for digital culture mediation], 2023. Accessed 4 November 2024
65. H. Needham, Evaluation of theatre, in *Case Studies in Museum*. ed. by L. Maloney, C. Hughes. Zoo and Aquarium Theater. (American Association of Museums, Washington, DC, 1999), pp.87–99

66. Y.-G. Noh, J.-H. Kong, Designing reenacted chatbots to enhance museum experience. Appl. Sci. **11**(7420), 1–15 (2021)
67. V. Okanovic, I. Ivkovic-Kihic, D. Boskovic, Interaction in extended reality applications for cultural heritage. Appl. Sci. **12**(3), 1241 (2022)
68. A. Pagano, A. Palombini, G. Bozzelli, M. De Nino, I. Cerato, S. Ricciardi, Arkaevision vr game: User experience research between real and virtual paestum. Appl. Sci. **10**(9), 3182 (2020)
69. R. Parry, *Recoding the Museum: Digital Heritage and the Technologies of Change* (Routledge, London, 2007)
70. M. Pasikowska-Schnass, Y.-S. Lim, Artificial intelligence in the context of cultural heritage and museums: Complex challenges and new opportunities. European Parliamentary Research Service, Brussels (2023). https://www.europarl.europa.eu/RegData/etudes/BRIE/2023/747120/EPRS_BRI(2023)747120_EN.pdf
71. S. Pearce, *Museums* (Objects and Collections. Smithsonian Books, Washington, DC, 1992)
72. A. Perla, Democratizing museum practice through oral history, digital storytelling, and collaborative ethical work. Santander Art and Culture Law Review **2**, 199–222 (2020)
73. S. Perry, M. Roussou, M. Economou, H. Young, L. Pujol, Moving beyond the virtual museum: Engaging visitors emotionally, in *2017 23rd International Conference on Virtual System & Multimedia (VSMM)*. (IEEE, 2017), pp. 1–8
74. P. Pierroux, M. Bäckström, B. Brenna, G. Gowlland, G. Ween, Museums as sites of participatory democracy and design, in *A History of Participation in Museums and Archives: Traversing Citizen Science and Citizen Humanities*. ed. by P. Hetland, P. Pierroux, L. Esborg (Routledge, New York, 2020)
75. E. Pietroni, Experience design, virtual reality and media hybridization for the digital communication inside museums. Applied System Innovation **2**(4), 35 (2019)
76. Th. Pistola, S. Diplaris, Ch. Stentoumis, E. Stathopoulos, G. Loupas, Th. Mandilaras, G. Kalantzis, I. Kalisperakis, A. Tellios, D. Zavraka, P. Koulali, V. Kriezi, V. Vraka, F. Venieri, S. Bacalis, St. Vrochidis, and I. Kompatsiaris. Creating immersive experiences based on intangible cultural heritage, in *IEEE International Conference on Intelligent Reality (ICIR)* (2021), pp. 17–24
77. A. Podara, D. Giomelakis, C. Nicolaou, M. Matsiola, R. Kotsakis, Digital storytelling in cultural heritage: Audience engagement in the interactive documentary new life. Sustainability **13**(3), 1193 (2021)
78. O. I. Psomadaki, C. A. Dimoulas, G. M. Kalliris, G. Paschalidis, Digital storytelling and audience engagement in cultural heritage management: A collaborative model based on the digital city of Thessaloniki. J. Cult. Heritage Manage. **36**, 12–22 (2019)
79. V. Polyzoi, F. Venieri, Engaging roma students with intangible heritage: Online live museum theatre at the museum of greek folk musical instruments, in M. Laraignée, editor, *Best Practices 12, A Tool to Improve Museum Education Internationally*, pages 78–90. International Council of Museums & International Committee for Education and Cultural Action (2024)
80. L. Pujol, M. Roussou, S. Poulou, O. Balet, M. Vayanou, Y. Ioannidis, Personalizing interactive digital storytelling in archaeological museums: The chess project, in *40th Annual Conference of Computer Applications and Quantitative Methods in Archaeology*. (Amsterdam University Press, 2012)
81. L. Pujol-Tost, Cultural presence in virtual archaeology: An exploratory analysis of factors. *Presence: Teleoperators and Virtual Environments*, 26(3):247–263 (2018)
82. H. Rahaman, *A Framework for Digital Heritage Interpretation*. PhD thesis, University of Singapore (2012)
83. H. Rahaman, Digital heritage interpretation: a conceptual framework. Digit. Creativity **29**(2–3), 208–234 (2018). https://doi.org/10.1080/14626268.2018.1511602
84. G. Raz, Rage against the empathy machine revisited: The ethics of the empathic affordances of virtual reality (2022)
85. S. Rizvic, D. Boskovic, V. Okanovic, Interactive digital storytelling: Bringing cultural heritage into the classroom. Journal of Computer Education **6**, 143–166 (2019)

86. S. Rizvic, N. Djapo, F. Alispahic, B. Hadjihalilovic, F. Fejzic Cenginc, A. Imamovic, Guidelines for interactive digital storytelling presentations of cultural heritage, in *2017 9th International Conference on Virtual Worlds and Games for Serious Applications (VS-Games).* (Athens, Greece, 2017), pp. 253–259
87. S. Rizvic, V. Okanovic, D. Boskovic, Digital storytelling, in *Visual Computing for Cultural Heritage*. ed. by F. Liarokapis, A. Voulodimos, N. Doulamis, A. Doulamis (Springer, Cham, 2020), pp.335–348
88. D. Rosenthal, Revisioning the city: Public history and locative digital media, in *Hidden Cities: Urban Space*. ed. by F. Nevola, D. Rosenthal, N. Terpstra. Geolocated Apps and Public History in Early Modern Europe. (Routledge, Oxon and New York, 2022), pp. 21–38
89. R. Rouse, Someone else's story: an ethical approach to interactive narrative design for cultural heritage, in *Interactive Storytelling: 12th International Conference on Interactive Digital Storytelling, ICIDS 2019, Little Cottonwood Canyon, UT, USA, November 19–22, 2019, Proceedings 12*. (Springer, 2019), pp. 47–60
90. M. Roussou, The components of engagement in virtual heritage environments, in *Proceedings of New Heritage: Beyond Verisimilitude - Conference on Cultural Heritage and New Media*. (Hong Kong, 2008), pp. 265–283
91. R. Rubenstein, H. Needham, Evaluation of the live interpretation program at the canadian museum of civilization, in *Perspectives on Museum Theatre*. ed. by C. Hughes (American Association of Museums, Washington, DC, 1993), pp.905–142
92. M.-L. Ryan, *Narrative Across Media: The Languages of Storytelling* (University of Nebraska Press, Lincoln, 2004)
93. E. Sansom, Peopling the past: Current practices in archaeological site interpretation, in *Archaeological Displays and the Public: Museology and Interpretation*. ed. by P. McManus (University College London, London, Institute of Archaeology, 2016), pp.113–118
94. E. Selmanović, S. Rizvic, C. Harvey, D. Boskovic, V. Hulusic, M. Chahin, Improving accessibility to intangible cultural heritage preservation using virtual reality. J. Comput. Cult. Heritage **13**(2) 1–19 (2020)
95. M. Shehade, T. Stylianou-Lambert (eds.), *Museums and Technologies of Presence* (Routledge, London, 2023)
96. N. Simon, *The Participatory Museum*. (Museum 2.0, Santa Cruz, 2010)
97. K. Stover, Is it real history yet?: An update on living history museums. J. Am. Cult. **12**(2), 13–17 (1989)
98. S. Sutherland Clothier, How theatrical is museum theatre? Museum Manage. Curatorship **29**(3), 211–225 (2014)
99. M. Taddeo, L. Floridi, Can ai be a force for good? Science **361**(6404), 751–752 (2018)
100. F. Tilden, *Interpreting Our Heritage* (University of North Carolina Press, Chapel Hill, NC, 1957)
101. G. Trichopoulos, G. Alexandridis, G. Caridakis, A survey on computational and emergent digital storytelling. Heritage **6**, 1227–1263 (2023)
102. Z. Tsiviltidou, G. Vavoula, Digital stories with the online collection of the v&a for inquiry-based learning. ACM Journal on Computing and Cultural Heritage **16**(1), 1–23 (2023)
103. F. Venieri, Facilitated dialogue: An emerging field of museum practice. *EXARC*, 2022/1, 2022. Accessed 5 Nov 2024
104. F. Venieri, Digital transition of museum theatre – an enrichment tool for history education, in European Union–Council of Europe joint project Transnational History Education and Cooperation Laboratory/HISTOLAB, editor, *Renewing History Education to Uphold Democracy*, pp. 63–90. Council of Europe, Strasbourg (2024)
105. F. Venieri, *Museum Theatre: History, Theory* (Implementation. Disigma, Thessaloniki, 2024)
106. F. Venieri, I. Marmaras, V. Tsetsos, Interactive museum theatre documentary: An innovative digital application for the research and interpretation of cultural heritage, in *Proceedings of the 3rd Panhellenic Conference on the Digitization of Cultural Heritage* (2019), pp. 259–268
107. F. Venieri, A. Gazi, Evaluating a digital heritage storytelling application: The case of revathens. Int. J. Inclusive Museum **18**(1), 209–232 (2024)

108. E. Vrettakis, V. Kourtis, A. Katifori, M. Karvounis, C. Lougiakis, Y. Ioannidis, Narralive: Creating and experiencing mobile digital storytelling in cultural heritage. *Digital Applications in Archaeology and Cultural Heritage*, 15 (2019)
109. A. Wong, The whole story, and then some: 'digital storytelling' in evolving museum practice. Museums and the Web, 2015. Accessed 5 Nov 2024
110. B. Wyman, S. Smith, D. Meyers, M. Godfrey, Digital storytelling in museums: Observations and best practices. Curator: Museum J. **54**(4), 461–468 (2011)

Chapter 3
Controlled Vocabulary Editor: SKOS Vocabularies without SKOS Training

Myrto Koukouli, Akrivi Katifori, Maria Roussou, and Yannis Ioannidis

Abstract The Semantic Web offers a unique opportunity to connect cultural heritage data and give rise to a global network of historical events and ideas that might change our view of history. Controlled vocabularies play a central role in this process. They serve as a common reference, a means for aligning the data. As linked data, they can even become a source of information themselves. Organizations of all sizes as well as individual researchers have been using them for a long time. However, the leap towards linked data presents challenges. One is the technical skills required to create or transform controlled vocabularies into linked data. The development of user-friendly applications addressed to users without technical expertise could be a solution. To this end, we have created and present in this chapter CoVEd, a user-friendly editor for controlled vocabularies as linked data in the SKOS schema. We evaluated the editor through a user study with 15 experts, aiming to record their perspective and expectations from such an application. The results suggest that an intuitive tool like CoVEd could facilitate experts to focus more towards the vocabulary and data and less towards the means to curate them, supporting the improvement of heritage data.

Keywords Controlled vocabularies · Semantic web cultural heritage · SKOS schema · Linked data

---

M. Koukouli (✉) · A. Katifori · M. Roussou · Y. Ioannidis
National and Kapodistrian University of Athens, Athens, Greece
e-mail: mkoukouli@di.uoa.gr

A. Katifori
e-mail: vivi@di.uoa.gr

M. Roussou
e-mail: mroussou@di.uoa.gr

Y. Ioannidis
e-mail: yannis@di.uoa.gr

K. Karpouzis and Y. Skarpelos (eds.), *Computational Methods for the Digital Humanities*, Learning and Analytics in Intelligent Systems 57,
https://doi.org/10.1007/978-3-032-05099-1_3

3.1 Introduction

Controlled vocabularies are predefined sets of terms or concepts that play an important role in many domains. Their use varies from organizing research work to classifying collections of thousands of assets or populating drop down lists in an application. They have been in use since long before the existence of computers and information systems. In the modern era they are one of the building blocks of the semantic web. For cultural heritage in particular, controlled vocabularies are the ingredient that makes the alignment of different datasets and collections possible [4, 10]. In a sense, they help unveil the hidden threads connecting cultural heritage beyond time and space. Because of the importance of vocabularies in cultural heritage, it is equally important to support their creation and maintenance. Since not all experts involved in their creation have technical skills, the development of user-friendly software to support them is crucial [14]. Controlled Vocabulary Editor (CoVEd) was created to fulfill this need while being an efficient editor for controlled vocabularies following the SKOS schema, the standard model for expressing them as linked data.

In the sections that follow an overview is given on the theoretical background and the application itself. The first section (Sect. 3.2) focuses on controlled vocabularies in general and the SKOS schema. In Sect. 3.3 the role of controlled vocabularies in cultural heritage is discussed and the requirements for a SKOS editor for cultural heritage. Section 3.4 makes reference to related work. The description of the application and its evaluation follow in Sects. 3.5 and 3.6. Finally, some conclusions and plans for the next steps in the development of the application are included in the last section (Sect. 3.7).

3.2 Controlled Vocabularies

Controlled vocabularies are Knowledge Organisation Systems used to consistently organize, classify or describe information. They often include variant terms logically classified, thus forming simple or more complex structures [4]. These relationships between the terms amplify the value of controlled vocabularies for a multitude of classifying and retrieval tasks and contribute to the discoverability of web resources and the facilitation of data exchange between organizations and datasets [10].

The size and complexity of controlled vocabularies vary and depend on the knowledge domains they cover and on the purpose of their creation [10]. Vocabularies that were created to provide a single reference for terms, such as the Library of Congress Subject Headings, are usually quite extensive and include several thousands of concepts, their scope, the relationships between them, etc. On the other hand, there are vocabularies that were created for specific applications or cover a limited domain and only need to include a handful of terms. Either way, their great value in knowledge organization and data interoperability remains the same.

3.2.1 Vocabulary Types

Controlled vocabularies can take several forms depending on the primary reason of their creation. Their structure can also vary from simple lists to complex directed graphs. The basic types of controlled vocabularies are [6, 12]):

Subject headings Consisting of words or phrases and their modifiers and used to describe concepts or subjects in general. A well-known example is the Library of Congress Subject Headings,[1] which has been in development since 1898.

Term lists Terminology listings created by organizations to serve their collections' specific needs. Such lists are maintained, for example, by regional research institutions and often include terms that correspond specifically to their collection.

Glossaries An alphabetical arrangement of terms accompanied by their definition, such as dictionaries.

Classifications A division of domain specific terminology into categories and subcategories while providing justification for this arrangement. An example is the ICONCLASS[2] vocabulary that was created to provide a universal classification system for works of art.

Taxonomies An arrangement of terms and concepts in groups and sub-groups according to their similarities.

Thesauri A collection of terms organized based on the semantic relationships between them to create a network of concepts. A very important thesaurus for the Cultural Heritage domain is the Art and Architecture Thesaurus[3] (AAT from now on).

3.2.2 As Linked Data

Controlled vocabularies play a significant part in today's information systems. Especially in the form of linked data they have moved beyond simple listings of terms. Each term now has a life of its own. It can be easily enriched with more information such as labels in many languages, notes on its scope, its history etc. and it can be linked to other resources [13]. It is now possible to define relations between vocabularies, thus facilitating their interoperability and reuse [7]. In this way, libraries and other organizations with pre existing vocabularies, for example, can achieve intercommunication while continuing to use essentially the same custom vocabularies [2]. Furthermore, controlled vocabularies as linked data can be easily made accessible online, thus multiplying their usage within scientific research while being available for use in tasks related to data-analysis and software applications development. They

[1] https://id.loc.gov/authorities/subjects.html.

[2] https://iconclass.org/.

[3] https://www.getty.edu/research/tools/vocabularies/aat/.

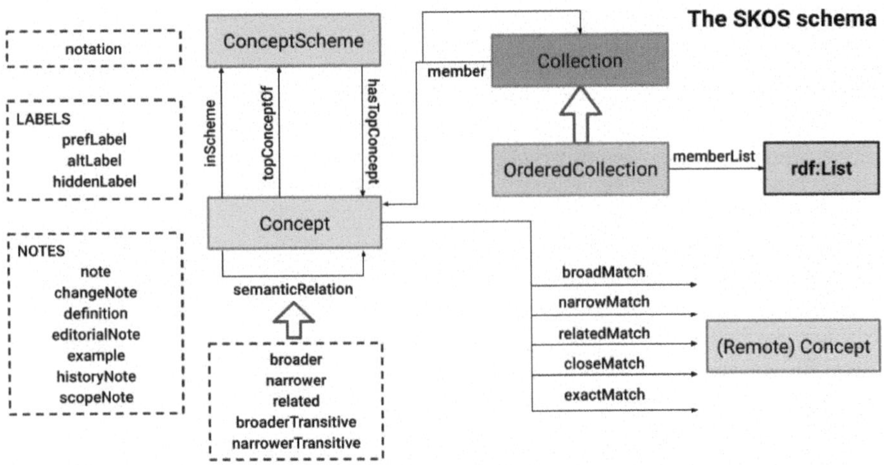

Fig. 3.1 The SKOS schema

even have the potential to become an important source of information themselves for a wide variety of end-users [13].

3.2.3 *The SKOS Schema*

The Simple Knowledge Organization System (SKOS) is a schema created to describe controlled vocabularies, especially ones with hierarchical or more complex structures such as thesauri and classifications, and to support their reuse in the Semantic Web [1]. It has been a W3C (World Wide Web Consortium) recommendation since 2009[4] and, as such, is widely used to express controlled vocabularies as linked data or simply to assign labels to resources. It is a simple schema consisting of four classes and twenty-five properties (Fig. 3.1). There also exists an extension of the core schema (SKOS-XL), a complementing set of one class and five properties that add more flexibility to the description of labels.[5] However, it is out of the scope of the present work. To make the next sections easier to follow, we have added a short description of the core SKOS schema, which is used by our application.

The basic building blocks of a SKOS vocabulary are the `Concept` and the `ConceptScheme`. The Concept class corresponds to the concepts/terms of the controlled vocabulary. The ConceptScheme class represents an individual vocabulary that contains a set of Concepts. Two additional classes, the `Collection` and the `OrderedCollection` serve the need to group Concepts according to some criterion. Each Concept can be linked to the ConceptScheme it belongs

[4] https://www.w3.org/TR/skos-reference/.

[5] https://www.w3.org/TR/skos-reference/skos-xl.html.

to via the `inScheme` property. The top, usually more general in meaning, concepts in the vocabulary hierarchy can be defined using the `hasTopConcept` and `topConceptOf` properties, the first is attributed to the ConceptScheme and the latter to the Concept. Membership of Concepts to the two collection classes of SKOS can be declared using the `member`, for both classes, and the `memberList`, only for the OrderedCollection class, properties.

Instances of the Concept class are interlinked by properties that represent semantic relations and ultimately form the hierarchical structure of the vocabulary and its connections to other SKOS vocabularies. The `broader` and `narrower` properties define the hierarchical relationships from one Concept to another one located immediately above it in the hierarchy of the same vocabulary and vice versa. The `broaderTransitive` and `narrowerTransitive` properties express the transitivity of the previous properties and are usually inferred based on the broader or narrower relationships that exist in the vocabulary. One additional property that represents a more generic relation between Concepts in the same vocabulary is the `related` property. Finally, a set of five properties define various levels of association between Concepts of different vocabularies. These are the `broadMatch`, `narrowMatch`, `relatedMatch`, `closeMatch` and `exactMatch` properties.

Other properties are used to clarify the characteristics or purpose of a Concept. The 'notation' property assigns an identifier to the Concept within the scope of the same ConceptScheme. Seven documentation properties (`note`, `changeNote`, `definition`, `editorialNote`, `example`, `historyNote` and `scopeNote`) facilitate the addition of further information on the context that the Concept represents as well as on details of its incorporation to the vocabulary.

Finally, there are three properties corresponding to types of labels, preferred labels (`prefLabel`), alternate labels (`altLabel`) and hidden labels (`hiddenLabel`). Preferred labels are limited to one per language. There are no restrictions for the other two. Due to their generic definition within the SKOS schema, these properties are also used in other data models, like the Europeana Data Model, as an agile way to label resources.

3.3 Controlled Vocabularies in Cultural Heritage

Cultural heritage (CH from now on) is a domain that has been using controlled vocabularies extensively. Big and small cultural organizations, individuals or groups of researchers, all utilize and maintain terminologies for types, materials, historical periods etc., which in many occasions are a result of their own research. They all could benefit greatly from the use of controlled vocabularies in the form of linked data, as a means to achieve interoperability among organizations and collections. The documentation of CH assets relies heavily on the observations and experience of the experts. It is thus more difficult than in other sciences to achieve a perfect alignment of data across institutions. Furthermore, because of the variations of human civilization in different regions or in different times in history, it is particularly challenging to decide on a single universal controlled vocabulary for all cases and switch to it [6]. Using controlled vocabularies as linked data does not necessarily require the direct application of a single, very analytic and complete vocabulary. It allows separate vocabularies to have common references and, as a result, to be interconnected by semantic links.

The ability to define relationships between terms allows institutions and researchers to retain their own particular terminology and still be able to produce universally aligned data by linking their terms to those of globally used vocabularies [6]. An important such global vocabulary for CH is the Art and Architecture Thesaurus (from now on, AAT), created and maintained by the Getty Institute. AAT contains an extensive collection of hierarchically arranged terms corresponding to most, if not all, aspects of human civilization [4]. Due to its extent and because it is recognised and used by institutions all over the world, AAT is a very important tool for the facilitation of the alignment of different controlled vocabularies. Institutions and researchers can create semantic links between the terms they use and terms from AAT. Even when an exact match is not found, a relationship to the closest term in the hierarchy can usually be drawn.

Other attributes of controlled vocabularies as linked data also benefit cultural heritage. Being able to manage translations, alternative names, definitions and other information on the terms is very useful. These labels and definitions of the terms can then be available to both the experts during the documentation process and the general public when visiting a digital collection.

3.3.1 Cultural Heritage in the Semantic Era

Since the emergence of linked data, extensive literature has been published on standards and good practices for the transformation of CH data in order to form part of the Semantic Web [7]. Controlled vocabularies play a crucial role in this process since the linking of CH data relies heavily on them. In recent years CH data in the Semantic Web have multiplied [14]. They are produced both by CH organizations of various sizes and by individual researchers [13]. The methods for assigning terms to the CH assets include automatically and semi-automatically annotating them using Natural Language Processing techniques [11] or manually choosing the terms.

While those facts seem promising in terms of the increased accessibility of CH collections, concerns have been raised about the difficulties that this transformation conceals and about the quality of the produced data. The challenges vary from inconsistency issues in the original data to licensing problems and from lacking trained staff to lacking user-friendly software [14]. These challenges in combination with the variety of strategies followed by different organizations lead to data quality issues and make interoperability difficult [7, 11].

A key element for quality of data in CH is the creation and use of controlled vocabularies [5]. The challenges of creating controlled vocabularies as linked data are similar to those that exist in general during the conversion of CH data to linked data [2]. For controlled vocabularies there is the added difficulty of ensuring the proper use of the terms by those who document the assets [4]. This issue was already present in legacy data. However, since they were meant to be read and studied by humans and not analyzed by machines, it didn't raise big concerns in the scientific community.

The involvement of more experts in the management of controlled vocabularies could potentially lead to better quality of data and the vocabularies' enrichment could enhance the presentation of CH digital collections to the public. Hierarchical and documented controlled vocabularies can be used for multiple purposes. The experts can have access to detailed terminology while visitors can use more general and understandable terms in order to search the collections in the organizations' web platforms. With linked data this can be achieved while maintaining the interoperability with other collections by creating strong links to common large-scale vocabularies [7].

3.3.2 Requirements for a SKOS Editor for Cultural Heritage

The present work aims to confront the challenges and issues described in the previous section by enabling more CH experts to get involved with the creation and maintenance of controlled vocabularies. CoVEd is a software that does not require prior knowledge of RDF[6] in order to create vocabularies as linked data, based on the SKOS schema. At the same time, it also aims to offer a fast way for users with knowledge of RDF to create a hierarchy of terms. The users can focus on documenting and organizing the terms they use in their research or in their organization and not in the technical details of the data formatting.

The specific requirements for the software can be summarized as follows: (1) The application should support the creation, management and view for both the concepts and the concept scheme. (2) The vocabulary must comply with the SKOS schema. (3) Navigation and visualization functions should be included and also, import and export options of vocabularies would be desirable. (4) Most importantly, the application should target a wide range of users and not only those that are familiar with RDF syntax. (5) Finally, the user interface should facilitate the quick editing of the vocabularies for both types of users.

3.4 Related Work

Many software applications have been developed for creating and maintaining controlled vocabularies. Some of them form part of a data management system or of a set of digital tools and others were implemented as individual applications. Furthermore,

[6] Resource Description Framework. The standard model for data interchange on the Web (https://www.w3.org/RDF/).

some of those applications produce lists or hierarchical structures of simple textual terms and some are used to create controlled vocabularies in the form of linked data. Since the SKOS schema is a standard, most applications follow this model.

As part of the development of CoVEd we researched the features and interface of twelve applications used to create SKOS vocabularies, as presented in the documentation and tutorials that are available. Most of the applications (ten of twelve) are web-based, one is a Windows OS application (Th Manager[7]) and one is a plugin[8] for the Protégé ontology editor.[9] The last two are now both discontinued, as is SKOS Editor, an application presented in [3]. Another one, Unilexicon,[10] is unclear if it is still maintained. Three are commercial products (TopBraid EDG,[11] PoolParty Taxonomy Manager[12] and Graphite[13]), all of which are part of a larger system. VocBench,[14] an open source and free software, is also part of a set of tools for managing rdf data. Additionally, it's the only one of the twelve applications that offer a separate, simpler, interface for managing vocabulary terms. However, as this, too, is part of a larger system oriented towards rdf data management in general, its interface seems rather complicated. Atramhasis[15] is a free open source web application dedicated to the management of controlled vocabularies. It has a simple interface and interesting features like importing concepts directly from other SKOS vocabularies but does not support the creation of a vocabulary from scratch. TemaTres[16] is a popular SKOS editor among the scientific community but seems to be targeting users that are already familiar with controlled vocabularies as linked data. iQvoc[17] is another free open source web application dedicated to creating and managing SKOS vocabularies. It has a user-friendly interface, includes a publishing option and integrates a user management system with separate user roles and a validation workflow leading to the publishing of the terms. Finally, Vocabs Editor,[18] also free and open source, has a simple interface many features for creating, editing and visualizing the vocabularies and, additionally, it integrates SKOSMOS,[19] a tool for browsing hierarchical vocabularies that makes the vocabularies accessible via a detailed RestAPI. Both iQvoc and Vocabs Editor are excellent examples of fully featured applications with user-friendly interfaces. They both limit the properties available during editing to

---

[7] https://thmanager.sourceforge.io/index.html.

[8] https://protegewiki.stanford.edu/wiki/SKOS_Editor.

[9] https://protege.stanford.edu/.

[10] https://unilexicon.com/vocabularies/.

[11] https://www.topquadrant.com/taxonomy-and-ontology-management.

[12] https://www.poolparty.biz/taxonomy-management-101.

[13] https://www.synaptica.com/graphiteknowledgestudio/.

[14] http://vocbench.uniroma2.it/doc/user/skos_editing.jsf.

[15] https://pypi.org/project/atramhasis/.

[16] https://vocabularyserver.com/web/.

[17] https://iqvoc.net/.

[18] https://vocabseditor.acdh.oeaw.ac.at/.

[19] https://skosmos.org/.

those of SKOS and a few others and, as far as we could understand, there is no way to use any others.

We wanted to experiment with the scenario of two types of users. The first ones should have the freedom to freely add triples to the ConceptScheme and the Concepts, as long as the SKOS schema is followed, and the others should be exposed to the actual schema and RDF syntax as little as possible. For these reasons we opted to continue and develop our own editor.

3.5 The Application

The COntrolled Vocabulary EDitor was created to serve two main purposes. First, to be a fast and easy-to-use SKOS editor that resembles writing directly in RDF, using the turtle syntax, and auto-generates some basic inferred triples. Second, to facilitate the creation of a SKOS vocabulary by users that have no knowledge of RDF or the SKOS schema. Our initial motivation was the need for a quick way to create hierarchical vocabularies for an on-going research project. The idea to make it approachable by a wider group of users came from the realization that experts of domains such as Cultural Heritage rely heavily on controlled vocabularies for their work but are rarely trained on RDF and Semantic Web technologies. In its current version CoVEd supports the SKOS schema, except for collections, but does not yet support the extended labels schema (SKOS-XL). Multiple concept schemes are also not supported and multiple broader relations in a single concept are supported only partially. The core characteristics and functionality of the application are briefly described in the sections that follow.

3.5.1 Technologies and Architecture

CoVEd consists of three parts: A database, a back-end and a user interface (Fig. 3.2). The data is stored in a graph database, GraphDB.[20] GraphDB was chosen because in addition to a SPARQL endpoint it offers a variety of API endpoints for managing sets of graphs (repositories), prefixes (namespaces) per set etc., as well as a built-in reasoner. The back-end is implemented in Python, using the Flask framework.[21] Finally, the user interface was built using the Angular framework[22] and communicates with the back-end via its RestAPI.

---

[20] https://www.ontotext.com/products/graphdb/.

[21] https://flask.palletsprojects.com.

[22] https://angular.io/.

Fig. 3.2 CoVEd architecture

3.5.2 *Functionality*

The application supports creating, editing and visualizing controlled vocabularies that are formed according to the SKOS schema. There are two available modes, a simple and an advanced one. The advanced version offers additional functionality, not available in the single mode. This section describes the full set of actions supported by the application. The user can create a new vocabulary from scratch or import one from a file. Creating a vocabulary actually creates the concept scheme for this vocabulary, which then serves as a single point of reference in all the created concepts. When a vocabulary is created or selected the user is redirected to the vocabulary management page (Fig. 3.3). This is where the development of the vocabulary takes place. Apart from the top and bottom bars, the page is divided in three sections vertically.

At the top section lie the prefixes used in the vocabulary. Prefixes can be added or removed at any time. If the vocabulary was imported from a file, the prefixes list is extended based on the prefixes used in it. Right below the prefixes there is the section for managing the concept scheme of the vocabulary (see Sect. 3.2.3 about the `ConceptScheme` and the `Concept` classes of the SKOS schema). Finally, the largest part of the screen is covered by the Concepts management section. Here the

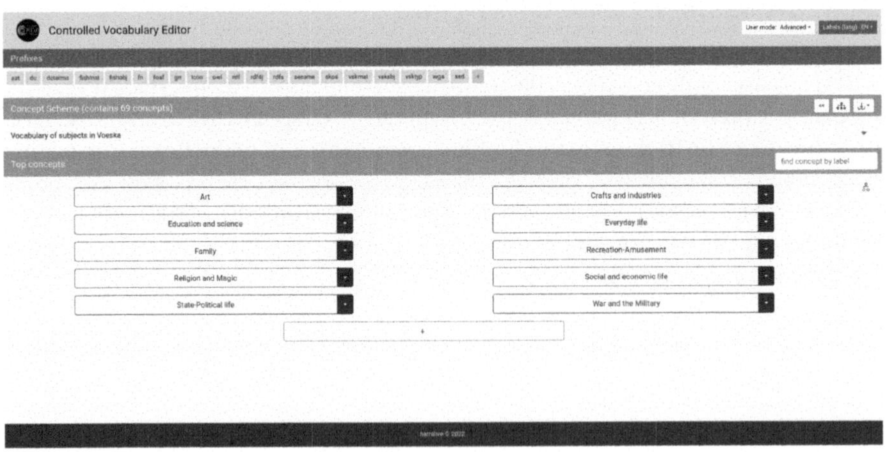

Fig. 3.3 The vocabulary management page in the advanced mode

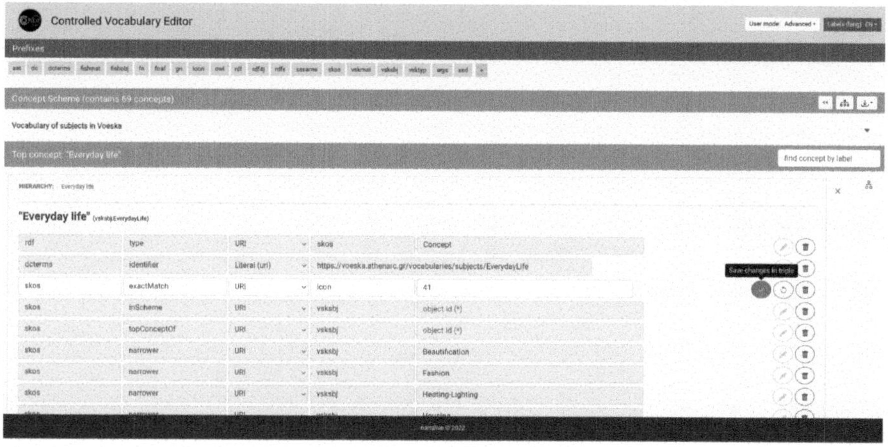

Fig. 3.4 The interface for editing a concept's triples

user can manage the triples of the concepts and the hierarchical relationships of the vocabulary.

The vocabulary is created gradually starting with the top concepts and moving to narrower concepts. Every time a new concept is created, the triples that define its location in the hierarchy are also added automatically. For example, when a top concept is created a `topConceptOf` and an `inScheme` triple are added to it and a new `hasTopConcept` to the concept scheme. Similarly, when a narrower concept is created a `broader` and an `inScheme` triple are automatically added to it, as well as a `narrower` triple to the concept under which it is added.

In edit mode, the triples of each concept or the concept scheme are edited separately and then saved as a set. To form a triple the user fills in a predicate and an object (Fig. 3.4). To speed things up, the properties defined in SKOS can be chosen from a dropdown list. To ensure the integrity of the hierarchy there are fail-safes that prevent the deletion of the semantic relations between the concepts. Additionally there are checks that ensure that other axioms and good practices are followed like having at least one preferred label, having at most one preferred label for a given language and not having duplicate preferred labels across the vocabulary.

In the view mode of the concepts, the triples that define relations are hyperlinks leading to the corresponding local or remote concept. This is one of the ways to navigate the vocabulary. There is also an option to search for a particular concept inside it by starting to type its label. One other way to navigate and get a closer look at the vocabulary is a horizontal tree diagram (Fig. 3.5) that can be opened on the side of the page and displays the currently selected concept and its immediate relations.

In addition to editing and viewing, there is an option to export the vocabulary as RDF, in turtle or xml format, or as an hierarchy of labels, in csv or MS Excel format. Lastly, the application includes one more page for viewing the vocabulary as a horizontal tree diagram. In this page the user can explore the vocabulary as a

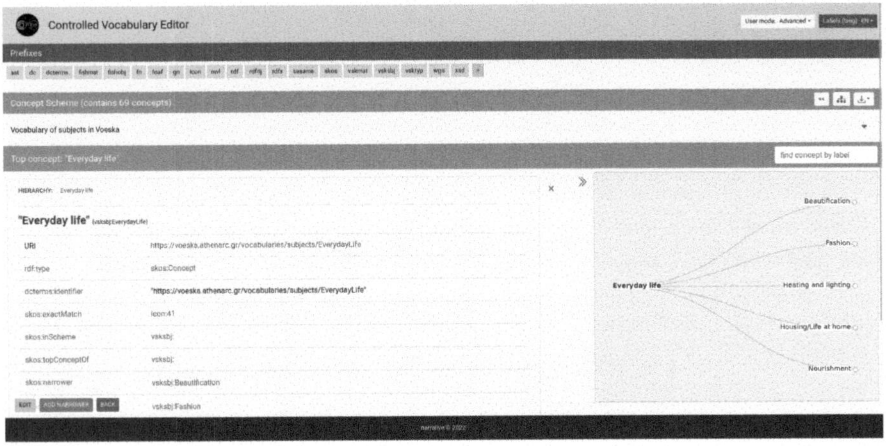

Fig. 3.5 The tree visualization of the current location in the vocabulary

whole, export the tree visualization as a png image or as a pdf file and navigate back to a specific concept in the previous page.

3.5.3 *The Simple and Advanced Modes*

Having worked alongside experts of cultural heritage and related disciplines, we understood the importance of developing an application that can be used regardless of one's knowledge of RDF, Linked Data or the SKOS schema. Since controlled vocabularies play a significant role in cultural heritage documentation, it stands to reason that an application for managing controlled vocabularies should target all possible users from this domain. The result of this thought is the development of both an advanced and a simple mode in the application.

The functionality remains more or less the same between the two modes with the exception of the prefixes management, which is only allowed in the advanced mode. In general, the simple mode offers specific predefined options whereas in the advanced mode the user has more freedom to customize the vocabulary. Starting from the creation of a vocabulary, in simple mode only a label is required whereas in advanced mode the user can add a custom prefix and uri. The option to import a vocabulary in rdf format from a file is available only in advanced mode. Moving to the vocabulary management page, for the creation of a concept in simple mode is again required just a label. In the advanced mode there is also an option to add a custom internal identifier. In the triples editing the user in simple mode is presented with the list of the labels of all SKOS properties (Fig. 3.6) and in advanced mode the process simulates writing them directly to turtle by filling in the prefix and id or

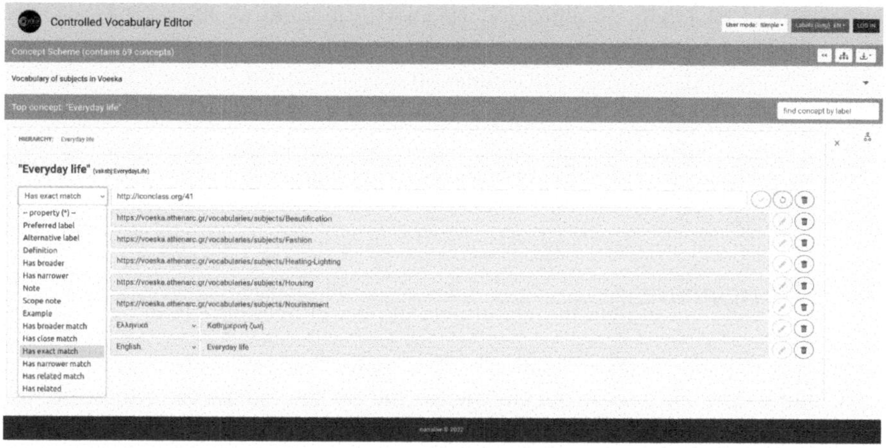

Fig. 3.6 The interface for editing a concept's triples in simple mode

the literal value of the predicate and object (Fig. 3.4). The only constraint is that for a prefix to be used it must first be added to the available prefixes.

The coexistence of the two modes offers an opportunity for groups of users to work together on developing their SKOS vocabulary even if only some or even none of the group members have knowledge of RDF and/or SKOS. On the other hand, it also demands more strict control of the possible modifications to the vocabulary, especially its hierarchy. The advanced user can freely use properties from schemata other than SKOS but the vocabulary must follow the basic SKOS structure for hierarchical vocabularies. It must have a concept scheme and top concepts under it and it must fulfill some basic rules like the uniqueness of preferred labels across a single vocabulary and the single preferred label for each language in a given concept [1]. In this way, the users can interchange the modes without risking changes that may not be supported in both. Since following proposed good practices is usually the best option, we consider this a reasonable trade-off. Furthermore, facilitating the involvement in the vocabulary development process of experts who will eventually use it, may be a significant step towards a better understanding of the concepts and their intended scope and, ultimately, towards higher quality of data.

3.6 Evaluation

The CoVEd design and development proceeded in iterative steps of formative evaluation and redesign with the support of users both with and without technical skills. This process led to the improvement of the usability aspects of the tool as well as of its functionality. In this chapter we present the evaluation of the beta version of the tool, focusing on the simple mode. Our objective was to assess the user experience aspects of the tool and its strengths and weaknesses in terms of offered functionality with a selected sample of users from a variety of fields closely related with Cultural Heritage To this end, we organized evaluation sessions where users would be invited for a hands-on assessment of the tool, guided by a series of tasks, while the evaluators observed their actions. The users were then asked to provide their feedback. In this section we present the evaluation process and its outcomes.

3.6.1 Participants

To invite participants for the evaluation we first approached experts from fields related with Cultural Heritage who, to our knowledge, did not have advanced technical skills. We asked them to take part themselves and/or forward the invitation to anyone else that they thought fit the profile. Fifteen experts accepted our invitation to participate in the evaluation.

According to the answers of the profile questions given to them, most of the participants (12 of 15) were between 25 and 44 years old and their specialty was related to the Cultural Heritage domain. The majority (11 of 15) had heard of Linked Data and the Semantic Web but had not seen or edited data in RDF form (10 of 15). About half of them had heard of Controlled Vocabularies and some (6 of 15) had used software for creating and editing controlled vocabularies again.

3.6.2 Process

As already mentioned, fifteen individuals participated in an equal number of evaluation sessions that were conducted remotely. The procedure was as follows: Each session started with a short presentation about the Semantic Web, Linked data and controlled vocabularies and an introduction to the main classes and properties of SKOS. A short video followed that gave the participants a glimpse of the application's basic functions, in the form of a tutorial. Next, the main part of the evaluation took place, during which the participant was given a set of tasks to complete using the application. The participants were asked to: create a vocabulary, add a second preferred label to its concept scheme in a different language, create a top concept, create a concept narrower to this top concept, navigate to another vocabulary and locate a concept relating to the one they created, return to CoVEd and add a mapping relation using the uri of the remote concept, test the navigation, search and export functions of the application on an already developed vocabulary. After the tasks were completed, the evaluator administered a semi structured interview and, finally, they completed an online evaluation questionnaire. This questionnaire contained:

- eight profile questions,
- the User Experience Questionnaire (UEQ),

- one question to measure the Net Promoter Score (a metric that is considered a standard for measuring customer loyalty and predicting future customer behavior) [8],
- an extra question about the usability of the application,
- five additional open-ended questions.

UEQ is a set of twenty six antithetic pairs of terms that are rated on a 7-point Likert scale and measure certain aspects of the user's experience of the application [9]. There is a dedicated website[23] for downloading the questionnaire in several languages and a tool that analyzes the responses, calculates the results and includes extensive explanation for the metrics and the aspects that are measured.

3.6.3 Results

The evaluation questionnaire started with some profile questions, the results for which are presented in Sect. 3.6.1 about the participants. The User experience questionnaire followed. The results for our application were encouraging. They can be viewed in Figs. 3.7 and 3.8. The UEQ benchmark (Fig. 3.7) measures the application performance in relation to six aspects of user experience: Attractiveness, Perspicuity, Efficiency, Dependability, Stimulation and Novelty. A second set of scales uses the measure for Attractiveness and groups the results for the rest of the aspects into two quality types (Fig. 3.8):

- Pragmatic quality (combines Perspicuity, Efficiency and Dependability and is related to the actual tasks that the application supports) and
- Hedonic quality (combines Stimulation and Novelty and has to do with non-task related aspects).

We can observe in Fig. 3.7 that the results fall well into the "Excellent" category for Attractiveness (the overall assessment), Stimulation (how much they enjoyed it) and, the highest, Efficiency (how quick it is to complete tasks) which was one of the principal motives for creating CoVEd. The results were also borderline Excellent for Perspicuity (how easy it is to learn to use it) and Dependability (how stable it feels). Last, the results fall into "Good" for novelty.

Following the UEQ, an additional usability question asked to the participants was whether they would use the application themselves or assign it to someone else. The majority (13 of 15) answered that they would use it themselves.

For the net promoter score we asked the participants how likely it would be for them to suggest the application to other people that use controlled vocabularies. The result showed that we have 14 promoters and 1 passive user. The net promoter score was 93, predicting that there is a high possibility our users would recommend the application to someone else.

---

[23] https://www.ueq-online.org/.

Scale	Mean	Comparisson to benchmark	Interpretation
Attractiveness	2.26	Excellent	In the range of the 10% best results
Perspicuity	2.07	Excellent	In the range of the 10% best results
Efficiency	2.35	Excellent	In the range of the 10% best results
Dependability	1.87	Excellent	In the range of the 10% best results
Stimulation	2.02	Excellent	In the range of the 10% best results
Novelty	1.45	Good	10% of results better, 75% of results worse

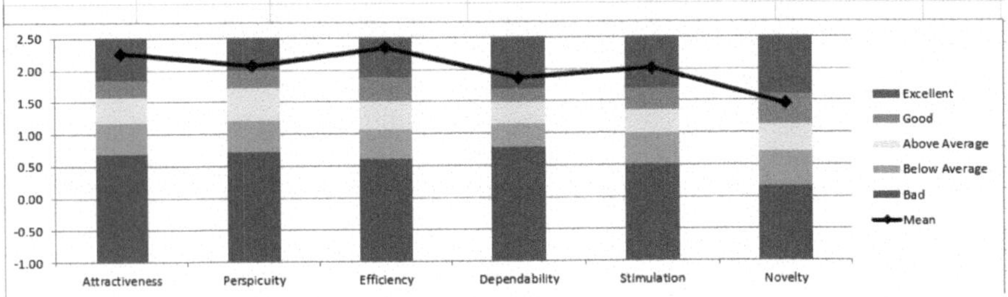

Fig. 3.7 Comparison with the UEQ benchmark *Source* screenshot from the analysis tool provided by the UEQ website

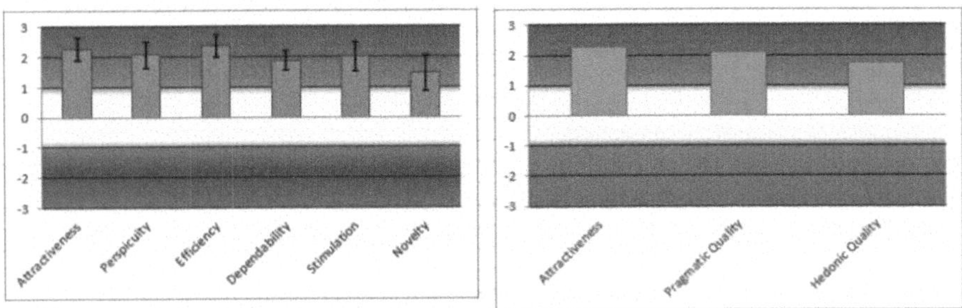

Fig. 3.8 UEQ results. *Source* charts copied from the analysis tool provided by the UEQ website

The results from the open-ended questions gave us interesting feedback on the participants' view of the application. The first such question was "To whom would this application be useful?" 8 of the participants mentioned CH specialists (archaeologists, historians, archivists, exhibition curators, collection managers and curators) and 6 mentioned researchers or scientists in general. The next most popular answers were experts working in data management (5) while 5 answers acknowledged usefulness in a wide range of domains. Librarians followed with 4 mentions. Finally, there was 1 answer that included school or university students to the potential users.

The next question was about difficulties that the participant may have faced during the evaluation process. No serious issues were mentioned, only minor difficulties that can be expected due to the lack of familiarity with the application.

In the question about suggesting changes and additions for the application the answers varied and many interesting features were proposed. An option to add images was suggested, further use of the tree diagram, the addition of a tutorial to assist first-time users, an option to import a vocabulary from file and to release a standalone version of the application.

The next question was "Which features of the application do you consider useful? Which not?". Here, the ability to create links to other vocabularies and the tree diagrams were recognised as the most useful features of the application with 4 mentions, 2 mentioned entering information in multiple languages. There was 1 mention for the ability to view and understand the relations between terms, for the search function and for the facilitation of enriching the terms with attributes. No one pointed out a not useful feature.

According to the final question of the questionnaire, "Which do you believe is the best aspect of the application?", the best aspect of the application with 6 mentions is the user interface, which was characterized as simple and understandable. The tree visualizations followed (3) and the ability to link to other concepts (2). Entering information in multiple languages and the aesthetics of the application had 1 mention each.

From the short interviews that we conducted after every evaluation session we came to the conclusion that most of the participants were not familiar with linked data or the Semantic Web, though some of them had heard of those terms. What was recognised by almost all of the participants was the usefulness of controlled vocabularies in several domains and the value of creating links to other vocabularies. They also made suggestions for additional attributes that could be added to each term (e.g. images) according to their idea of the task for which they would use the application. A most interesting fact is that almost every user offered his/her own distinct suggestion for using the application. Some recognized its value for education and even made a plan about how they could use it to create concept trees and present them to their students. Others were excited about adding their own notes and references in the terminology they use in research. The few more advanced users argued that it could be used to easily create meta-vocabularies or as a simple way to maintain a vocabulary and detect inconsistencies.

The results of the evaluation were very encouraging as the users welcomed the tool with enthusiasm and made several suggestions towards its improvement. The activity confirmed that the tool can be used effectively by users with no particular technical expertise, even with minimum training and without knowledge of RDF.

Following the evaluation, some of the suggestions have already been adopted, such as importing vocabularies from a file. Also, steps towards further simplifying the user interface have been taken.

3.7 Conclusions and Next Steps

The Semantic Web has made the intercommunication and data exchange between cultural heritage institutions more effective. Controlled vocabularies play a crucial part in this web of data that keeps growing at a high speed. Projects like the AAT provide a powerful means for this interlinking to expand. However, the development and maintenance of custom controlled vocabularies that refer to the AAT but serve the specific needs of an organization, a researcher or a small group of experts working together on a project is still relevant. Similar needs surface when integrating a vocabulary to a software application. Issues like the translation of labels to local languages, adding custom definitions and scope notes to the concepts and adding references to preexisting and possibly long used vocabularies.

CoVEd aims to make the creation of SKOS vocabularies easier for those that may have the need for a customized controlled vocabulary. The range of users can be as wide as is the use of controlled vocabularies. What our application offers is the opportunity to effortlessly modify and annotate the terminology one uses and be able to export it as linked data, ready to use. Alternatively, CoVEd can simply be used as a quick way to keep at the same place the terms you use during the documentation of assets and the notes and information on what those terms mean and how they are used in your work. For groups of experts this can serve as a single reference for the use of terminology and, as a result, an easier way to achieve homogeneous documentation for the assets in their collection.

Designing a simple mode was not without challenges but, in our opinion, it was necessary in order to approach all the users who are directly involved in the development or the use of vocabularies. The results of the evaluation showed us that our first attempt was well received and the feedback from the participants was invaluable. More evaluation sessions should follow including also advanced users whose advice may lead to important improvements and extensions.

For the next steps in CoVEd's development, we plan to extend its functionality to support extended labels, more complex vocabulary structures, and SKOS Collections. We are also considering implementing more of the suggestions of our test users, such as adding a lookup function to find matching concepts from widely used select vocabularies, such as the AAT, to make it easier for users to link them to their own.

To sum up, it is advisable for those who use the vocabularies to have a good understanding of the concepts inside them [4]. Their involvement in the development of the vocabularies should provide them with a better perspective over their scope. A tool like CoVEd could facilitate the involvement of all interested parties without them having to learn RDF. In general, more user-friendly software applications could turn the focus more towards the content itself and less towards the means to create it and could potentially lead to an improvement in the quality of the produced data.

Acknowledgements This chapter reports on the outcomes of Myrto Koukouli's maste'sr thesis. I would like to thank Dr. Akrivi Katifori, Prof. Maria Roussou and my supervisor, Prof. Yannis Ioannidis, who supported me every step of the way. Thank you!

References

1. T. Baker, S. Bechhofer, A. Isaac, A. Miles, G. Schreiber, E. Summers, Key choices in the design of Simple Knowledge Organization System (SKOS). J. Web Semant. **20**, 35–49 (2013). https://doi.org/10.1016/j.websem.2013.05.001
2. E.R. Barbosa, M.L. Dutra, A.F. Godoy Viera, D.D.J.d. Macedo, Thesaurus and subject heading lists as Linked Data. Transinformação **33**, e200077 (2021). https://doi.org/10.1590/2318-0889202133e200077

3. M. Conway, A. Khojoyan, F. Fana, W. Scuba, M. Castine, D. Mowery, W. Chapman, S. Jupp, Developing a web-based SKOS editor. J. Biomed. Semant. **7**(1), 5 (2016). https://doi.org/10.1186/s13326-015-0043-z
4. P. Harpring, *Introduction to controlled vocabularies terminology for art, architecture, and other cultural works*. (Getty Research Institute, Los Angeles, California, 2010). http://www.getty.edu/publications/virtuallibrary/160606018X.html
5. M. Hlava, B. Kasenchak, Smart Thesauri: Using Taxonomies with Linked Data (follow up paper), in *Innovations in Knowledge Organisation Conference (IKO) 2015*. (Singapore, 2015). http://www.greenchameleon.com/uploads/Margie_Bob_v1.pdf
6. E. Lanzi, H. Besser, *Introduction to vocabularies: enhancing access to cultural heritage information* (Getty Information Institute, Los Angeles, Calif, 1998)
7. K. May, C. Binding, D. Tudhope, Barriers and opportunities for Linked Open Data use in archaeology and cultural heritage. Archaeol. Inf. **38** (2015). https://doi.org/10.11588/AI.2015.1.26162
8. Project Management Institute (ed.): The standard for project management and a guide to the project management body of knowledge (PMBOK guide), seventh edition edn. Project Management Institute, Inc, Newtown Square, Pennsylvania (2021)
9. M. Schrepp, A. Hinderks, J. Thomaschewski, Construction of a Benchmark for the User Experience Questionnaire (UEQ). Int. J. Interact. Multimedia Artif. Intell. **4**(4), 40 (2017). https://doi.org/10.9781/ijimai.2017.445, http://www.ijimai.org/journal/node/1514
10. C. Smith, Controlled vocabularies: Past, present and future of subject access. Catal. Classification Quart. **59**(2–3), 186–202 (2021). https://doi.org/10.1080/01639374.2021.1881007
11. G. Sugimoto, Instance level analysis on linked open data connectivity for cultural heritage entity linking and data integration. Semantic Web **14**(1), 55–100 (2022). https://doi.org/10.3233/SW-223026
12. Zaytseva, K.: Controlled Vocabularies and SKOS. DARIAH-Campus (2020). https://campus.dariah.eu/en/resource/posts/controlled-vocabularies-and-skos
13. M.L. Zeng, P. Mayr, Knowledge organization systems (kos) in the semantic web: a multi-dimensional review. Int. J. Digit. Libr. **20**(3), 209–230 (2018). https://doi.org/10.1007/s00799-018-0241-2
14. L. Zhang, Empowering linked data in cultural heritage institutions: A knowledge management perspective. Data Inf. Manage. **6**(3), 100013 (2022). https://doi.org/10.1016/j.dim.2022.100013

Chapter 4
AI, Digital Humanities, and the Legacies of Colonial Power

Kostas Karpouzis

Abstract This paper examines the impact of artificial intelligence (AI) on Digital Humanities through a colonial lens, analyzing how AI can both reinforce and challenge colonial power dynamics. AI tools in Digital Humanities, such as text mining and language preservation, often perpetuate Western epistemologies and marginalize non-Western perspectives due to biases in data and algorithms. Using case studies, such as the Slave Voyages database and indigenous language preservation projects, this paper highlights AI's dual role as both a potential perpetuator of colonial legacies and a tool for decolonization. It recommends inclusive AI design, community-driven data governance, and the integration of alternative epistemologies to mitigate AI's colonial biases and promote more equitable knowledge production.

Keywords Artificial intelligence · Digital humanities · Colonialism · Bias · Natural language processing

4.1 Introduction

In recent years, the intersection of Artificial Intelligence (AI) and Digital Humanities (DH) has opened new pathways for analyzing, interpreting, and preserving historical and cultural data. AI techniques, such as machine learning, natural language processing (NLP), and computer vision, have allowed researchers to tackle vast datasets, ranging from digitized texts and images to archives and metadata, with unprecedented speed and precision. These innovations are transforming the field of Digital Humanities by providing tools to uncover patterns, identify relationships, and make inferences that were previously impossible to discern manually. However, the integra-

K. Karpouzis (✉)
Department of Communication, Media and Culture, Panteion University of Social and Political Sciences, Athens, Greece
e-mail: kkarpou@panteion.gr
URL: https://kkarpou.github.io/

K. Karpouzis and Y. Skarpelos (eds.), *Computational Methods for the Digital Humanities*, Learning and Analytics in Intelligent Systems 57,
https://doi.org/10.1007/978-3-032-05099-1_4

tion of AI into the humanities also brings forth critical questions, particularly when viewed through the lens of colonial power dynamics and knowledge production.

From the point of view of Digital Humanities, related projects often focus on processing and analyzing cultural heritage data, including historical texts, artifacts, and archives. AI techniques have proven useful for projects such as digitizing and categorizing vast colonial archives, reconstructing lost languages, or creating interactive historical maps. These applications hold great potential for advancing research and public understanding of global histories and cultures. However, as these technologies expand, it becomes crucial to scrutinize how AI-driven methodologies might reproduce colonial frameworks in their approach to data collection, analysis, and representation.

The power of AI to influence Digital Humanities lies not only in the scale of analysis but also in the underlying assumptions encoded in algorithms. Many AI systems are trained on datasets predominantly sourced from Western contexts, with methodologies rooted in Eurocentric epistemologies. These frameworks risk perpetuating the dominance of colonial narratives by privileging Western data and ways of knowing. Consequently, AI may unintentionally reinforce the very colonial power structures that Digital Humanities scholars seek to deconstruct.

This colonial perspective highlights the uneven distribution of knowledge and power that emerged from centuries of colonial rule. Scholars such as Risam [19], Chakrabarty [4], and Benjamin [2] have critiqued how colonialism shaped the production and dissemination of knowledge, marginalizing non-Western voices and perspectives in the process. This dynamic persists in the digital realm, as digital archives, AI tools, and even the infrastructure behind these technologies are disproportionately controlled by the Global North. This has led to the rise of terms like digital colonialism and data colonialism, which emphasize the continuation of exploitative practices in the digital age.

For example, archives from colonial periods, when digitized and processed by AI, often reflect the biases of the original curators–who were often colonial administrators. AI systems, unless carefully curated, risk replicating these biases by amplifying the voices and narratives of colonial powers, while silencing indigenous and marginalized groups. This calls into question the neutrality of AI as a tool and urges a critical examination of its role in shaping how historical and cultural narratives are produced and understood [5].

In this paper, we will explore the following central questions:

- How do AI-based techniques in Digital Humanities reproduce or challenge colonial power structures?
- In what ways do AI-driven projects in Digital Humanities favor Western epistemologies and exclude marginalized or indigenous perspectives?
- How can AI be employed as a tool for decolonizing Digital Humanities, ensuring that non-Western voices are included in global knowledge production?

The next sections discuss the theoretical framework of AI influence and the related biases, present a number of examples of how this influence is manifested in action, and propose techniques to mitigate this effect.

4.2 Theoretical Framework

AI is proving to be a transformative technology in Digital Humanities, involving a range of computational techniques designed to mimic human or augment intelligence. The most prominent AI methods applied in this domain include:

- **Machine Learning (ML)**: A subset of AI, machine learning involves training algorithms on large datasets to identify patterns, classify information, and make predictions without explicit programming. ML in Digital Humanities can assist in analyzing cultural artifacts by recognizing patterns across large datasets, such as identifying trends in historical texts, categorizing images, or clustering themes across vast archives. Common techniques in ML include supervised learning (where models are trained on labeled data), unsupervised learning (pattern detection without labels), and deep learning (which uses neural networks for more complex pattern recognition).
- **Natural Language Processing (NLP)**: NLP refers to AI's ability to understand, interpret, and generate human language. This is particularly relevant in the analysis of historical documents, literature, and digitized texts. NLP is used in tasks like text mining, sentiment analysis, and automated translation. Projects involving the digitization of colonial archives often apply NLP techniques to mine vast quantities of text, identifying recurring themes or uncovering previously hidden connections within historical records.
- **Computer Vision**: AI's computer vision capabilities allow machines to interpret and categorize visual information, making this technology especially useful for analyzing and preserving visual artifacts like paintings, photographs, and sculptures [16]. In Digital Humanities, computer vision can be applied to recognize patterns in art history, detect visual elements in colonial-era photographs, or even reconstruct incomplete artifacts.
- **Generative AI**: Generative models, like Generative Adversarial Networks (GANs), are increasingly being used in the humanities to generate art or text based on learned data. For example, these models can be trained on a dataset of colonial-era art or texts and then generate new, derivative works that reflect historical styles or content [8].

While AI offers tremendous potential for Digital Humanities, it also presents challenges. Many of these techniques depend heavily on the quality and composition of the data they are trained on, which can lead to biased results if not carefully curated. In the context of colonial history, this bias may manifest as the reinforcement of Eurocentric narratives or the exclusion of indigenous perspectives, making it essential to approach AI's use critically and ethically in this field [12].

4.2.1 Colonialism: Definition and Framework

Colonialism refers to the political, social, and economic domination of one group over another, typically by a state or empire. Colonialism is not only a historical phenomenon but also an intellectual and cultural practice that extends beyond the physical occupation of territories. At its core, colonialism involves the extraction of resources, labor, and knowledge from colonized lands and peoples for the benefit of the colonizer.

The coloniality of power is a concept developed by sociologist Aníbal Quijano [18] to describe the lasting effects of colonialism on global systems of power. He argues that colonialism did not end with the formal independence of colonized nations but persists in the form of global hierarchies that privilege European modes of knowledge, economic systems, and social structures. This power dynamic is often reinforced through education, law, and media, where Western knowledge systems are seen as superior and universal. Colonialism also deeply impacted knowledge production. As postcolonial scholars like Said [20] have argued, colonial powers imposed their own ways of understanding the world on the peoples they colonized, marginalizing indigenous knowledge systems and languages. This legacy persists in modern academia and in many of the digital projects that aim to preserve or analyze historical artifacts, where Western epistemologies are often privileged at the expense of non-Western ones.

In the context of this work, the rise of AI and digital technologies has introduced new forms of colonialism, often described as data colonialism and digital colonialism. These terms describe the exploitation of data and digital infrastructures by powerful entities, often based in the Global North, to exert control over information, resources, and even people in the Global South.

The concept of *Data Colonialism* was developed by Nick Couldry and Ulises Mejias [6], and refers to the ways in which data is harvested from individuals and communities–often without their knowledge or consent–and then used for profit or control by large corporations and governments. The collection of data from the Global South by entities in the Global North mirrors historical patterns of resource extraction during colonial times. For example, many AI systems require vast amounts of data to function effectively, and this data is often sourced from populations in developing countries without their explicit consent or meaningful compensation. This creates a new form of dependency, where individuals and communities in the Global South become data subjects, while the benefits of AI systems are concentrated in the hands of a few powerful actors.

A related concept, *Digital Colonialism*, refers to the broader control of digital infrastructures, platforms, and technologies by companies and governments in the Global North. As with traditional colonialism, this involves the extraction of value–in this case, data and digital labor–by entities that have disproportionate control over the digital tools and platforms used globally. Scholars argue that digital colonialism perpetuates existing power imbalances, as countries and communities in the Global South are often forced to rely on digital infrastructures (such as cloud services, social media platforms, and AI tools) that are owned and controlled by foreign entities. This creates an unequal distribution of power and control over digital knowledge and resources.

These forms of colonialism are particularly relevant in the context of AI-driven projects in Digital Humanities. For instance, many digital archives of colonial history are housed in institutions in Europe and North America, where they are digitized and analyzed using AI tools. The narratives that emerge from these projects are often shaped by Western epistemologies and priorities, rather than by the perspectives of the people whose histories are being analyzed. This raises important ethical questions about who controls the past, who benefits from the digitization of cultural heritage, and how AI might be used to either reinforce or challenge these power dynamics.

4.2.2 Colonialism in Action

In this section, we explore how AI-driven projects in Digital Humanities reflect or challenge colonial legacies. By analyzing specific examples, we can first highlight the role AI plays in either reinforcing Eurocentric knowledge systems, before proceeding to investigate opportunities for decolonization and inclusivity.

One prominent application of AI in Digital Humanities involves the digitization and analysis of colonial archives. For example, large-scale projects that focus on colonial-era texts, legal documents, and administrative records have become a fertile ground for AI's text-mining capabilities. However, without critical reflection, these projects can reinforce colonial perspectives by foregrounding the narratives of colonial powers while marginalizing indigenous voices.

The Slave Voyages Database One well-known example of AI-driven Digital Humanities is the Slave Voyages project[1] This digital archive uses AI and data analysis techniques to map transatlantic slave trade routes, analyzing records from over 36,000 voyages that transported enslaved Africans [9]. While the project offers important insights into the scale and mechanics of the transatlantic slave trade, it also raises critical questions about representation and the limits of data. The archive relies heavily on colonial records, often written by the enslavers themselves, and while it includes some data on African perspectives, it inherently prioritizes the colonizers' viewpoint, since the records were produced by those in power.

Colonial Analysis: This case reveals how AI-driven Digital Humanities projects, when built on colonial archives, can unintentionally perpetuate colonial worldviews. The data used in the Slave Voyages project is derived from a system that documented people as property, reflecting the dehumanizing logic of the time. As such, while the database offers an important historical resource, it also risks erasing the agency and

[1] Slave Voyages: The Trans-Atlantic Slave Trade Database, Emory University, https://www.slavevoyages.org.

narratives of the enslaved unless complemented with efforts to include oral histories, indigenous archives, and other non-colonial perspectives.

Google's AI for Indigenous Language Preservation AI has also been applied in efforts to preserve endangered languages, many of which are spoken by indigenous communities. For example, machine learning and NLP technologies have been used to digitize and revitalize languages that are on the brink of extinction, thereby offering a potential tool for decolonizing knowledge production. However, these projects must navigate ethical challenges, particularly around ownership, consent, and the right to control indigenous knowledge.

Google's AI initiatives to preserve endangered languages, such as the use of its platform to digitize and document indigenous languages, are framed as decolonial efforts aimed at cultural preservation. However, critical voices have pointed out that this form of digitization can lead to new forms of appropriation and control, where the ownership of indigenous knowledge is shifted to global tech corporations. Indigenous communities have expressed concern over who owns the data once digitized, and whether they retain control over how their language and cultural knowledge are used.

Colonial Analysis: While AI offers powerful tools for language preservation, projects like these can fall into the trap of digital colonialism if indigenous communities are not actively involved in the process. The act of digitizing these languages can replicate the same extractive processes that defined colonialism, where the knowledge of colonized peoples was appropriated and commodified by colonial powers [22].

Mapping Colonial India through Image Recognition Computer vision has been employed in projects aimed at analyzing colonial-era artworks and photographs. While this technology allows for the categorization and preservation of visual materials, it often reinforces Eurocentric notions of art and cultural heritage by prioritizing Western classification systems and neglecting non-Western art forms.

An AI project at Oxford University sought to use computer vision to analyze photographs from colonial India. The AI system was trained to recognize and categorize images based on a dataset composed of British colonial records and photographs. While the project provided valuable insights into British colonialism's visual culture, it also raised ethical concerns. By focusing on colonial photographs, the AI system reinforced a specific narrative of India that prioritized the colonizers' gaze and treated Indian subjects as passive objects of the British Empire's control.

Colonial Analysis: This project demonstrates how AI, when trained on colonial-era data, can perpetuate colonial worldviews. By prioritizing British colonial records, the AI system reinforced a perspective that depicted Indian subjects through the lens of British control [17], rather than offering a more nuanced or indigenous understanding of these images.

The Colonial Despatches Project Digital archives are a central focus of AI applications in Digital Humanities, offering a wealth of material for text mining and analysis. However, many of these archives consist of colonial records that reflect the biases of their creators. Without critical intervention, AI tools that analyze these archives may unintentionally reinforce the same colonial ideologies that shaped the original records.

The Colonial Despatches project[2] led by scholars in Canada, digitized and made publicly available a collection of despatches (official communications) from British colonial administrators to the British government during the 19th century. The project applied AI techniques such as text mining to analyze the language used by colonial officials in their governance of indigenous populations in British Columbia. While the project has contributed significantly to understanding colonial governance, it relies heavily on the voices of colonial administrators, often excluding the perspectives of the indigenous peoples affected by these policies.

Colonial Analysis: This project raises important questions about whose voices are privileged in the historical record. While AI has made it possible to process and analyze these texts at scale, the focus on official colonial documents risks perpetuating a one-sided view of history [3]. Projects like these highlight the need to complement AI-driven archival research with efforts to include indigenous oral histories and perspectives.

These case studies illustrate the dual potential of AI in Digital Humanities: while AI can, indeed, uncover new insights into historical data, it can also reinforce colonial narratives if not applied critically. Each of the above mentioned projects reveals how AI's reliance on colonial-era data or Western epistemologies can perpetuate colonial power dynamics, as discussed in the next section. However, with thoughtful design, AI can also provide a platform for decolonizing knowledge, amplifying marginalized voices, and challenging the dominance of Eurocentric narratives in the digital sphere.

4.3 AI's Role in Reinforcing Colonial Structures

Artificial Intelligence is often seen as a neutral or objective technology, but its deployment discussed above shows that it can actually reinforce colonial structures through algorithmic bias, data imbalances, and the privileging of Western epistemologies. As a result, AI systems, in their design, training, and deployment, can perpetuate historical injustices by embedding colonial perspectives into their outputs.

One of the main means to this is through algorithmic bias: this arises when AI systems–trained on datasets that reflect existing social, political, and historical biases–reproduce these biases in their outputs. In the context of Digital Humanities, this bias often results in the privileging of Western knowledge systems at the

[2] Colonial Despatches: The Colonial Despatches of Vancouver Island and British Columbia 1846-1871, University of Victoria, https://www.colonialdespatches.ca.

expense of non-Western or marginalized perspectives. AI's dependence on large, pre-existing datasets, many of which are sourced from historically dominant (often Western) institutions, means that these systems are inherently biased toward the values, languages, and worldviews embedded in those datasets. An example of this is evident in NLP (Natural Language Processing) systems, which are frequently used in Digital Humanities for text mining and analysis, are a prime example of how algorithmic bias can manifest. These systems are often trained on English or other dominant languages, which means they are less effective at processing texts written in non-Western or indigenous languages. Even when NLP systems are applied to digitized historical texts from colonized regions, the interpretation of these texts is often influenced by the linguistic and cultural biases of the training data, which tends to reflect Western norms and values [15].

AI systems in Digital Humanities often reinforce what philosopher Miranda Fricker calls *epistemic injustice* [10], that is the systematic exclusion of certain groups from knowledge production. In colonial contexts, this means that the perspectives of indigenous or colonized peoples are often marginalized or erased, while the narratives of colonial powers are privileged. AI systems trained on colonial records are likely to perpetuate this dynamic unless deliberately designed to include alternative epistemologies or to counterbalance historical biases.

Besides this, AI's dependence on vast quantities of data raises critical issues related to data colonialism [5] to describe how digital technologies are used to extract and commodify data from individuals and communities–particularly in the Global South–without equitable benefit-sharing. In Digital Humanities, data colonialism manifests in the digitization of colonial archives and cultural artifacts, where the digital representation of these materials often ends up controlled by institutions in the Global North. In this context, large AI-driven projects that digitize colonial archives–such as the Endangered Archives Programme[3] at the British Library–are often framed as efforts to preserve global cultural heritage. However, critics have pointed out that these digitization projects can reinforce the control of Western institutions over cultural artifacts from the Global South. While the digital versions of these artifacts may be made publicly available, they remain housed in servers and systems owned by institutions in the Global North, perpetuating a colonial dynamic where Western entities maintain ownership and control over the knowledge and cultural production of formerly colonized peoples. This reflects a broader pattern of digital extraction, where the data and cultural resources of the Global South are harvested by corporations and institutions based in the Global North, mirroring historical patterns of resource extraction during colonial times, where the wealth of colonized nations was transferred to Europe and other colonial powers [1].

Another example of reinforcement of Western Epistemologies is manifested when AI systems in Digital Humanities rely on classification schemes, taxonomies, and ontologies that reflect Western academic traditions. These frameworks, embedded in AI algorithms, privilege certain ways of organizing knowledge and marginalize alternative epistemologies. For example, the way historical records are categorized

[3] Endangered Archives Programme, https://eap.bl.uk/.

and interpreted through AI tools often mirrors Western historiographical methods, which tend to emphasize linearity, causality, and written documentation–modes of historical inquiry that may not align with the traditions of oral or non-linear histories common in many non-Western cultures [19].

An example of this is AI-based analysis of colonial archives. Many AI systems are designed to categorize and interpret historical data according to Western scholarly traditions, which can lead to the exclusion of indigenous or alternative historical narratives [14]. Indigenous histories, which often prioritize oral traditions and cyclical understandings of time, may be poorly represented in AI-driven projects because these systems are built on Western ontological frameworks that emphasize written documentation and linear time. AI systems that rely on these frameworks perpetuate a form of epistemic dominance, where Western ways of knowing are treated as universal or superior. This leads to the marginalization of non-Western epistemologies and reinforces the historical silencing of colonized peoples.

4.4 Decolonizing AI in Digital Humanities

In this section, we explore how AI can be employed to address and mitigate the colonial legacies discussed earlier. Decolonizing AI in Digital Humanities involves reshaping AI systems to be more inclusive of marginalized perspectives, ensuring ethical data use, and incorporating non-Western epistemologies into AI frameworks. This process requires the collaboration of technologists, humanists, and marginalized communities to redefine the goals and methods of AI applications in Digital Humanities.

4.4.1 Practical Examples

One of the first steps toward this is ensuring that the voices and knowledge systems of indigenous and marginalized communities are represented in the data and algorithms used. This can be achieved by working directly with these communities to curate datasets and develop AI models that reflect their cultural values, histories, and languages. An example project is the *Algonquian Linguistic Atlas*[4] a collaborative Digital Humanities project that uses AI to help preserve and revitalize the Algonquian languages spoken by indigenous groups in North America. By working closely with indigenous communities, linguists, and AI developers, the project has created a platform where speakers can record and share their languages, ensuring that these languages are preserved in a way that honors their cultural significance [11]. This project highlights how AI can be used to decolonize knowledge production by prioritizing indigenous agency and control over the representation and use of their

[4] Algonquian Linguistic Atlas, https://www.atlas-ling.ca.

cultural data. Unlike many Western-centric projects, the Algonquian Linguistic Atlas allows the communities themselves to determine how their language and culture are represented digitally.

Another de-colonization opportunity stems from the fact that decolonizing AI requires the inclusion of alternative epistemologies–ways of knowing that do not conform to Eurocentric frameworks. This includes recognizing the importance of oral histories, communal memory, and non-linear conceptions of time, all of which are integral to many non-Western cultures. AI tools can be adapted to respect these alternative epistemologies by incorporating ethical guidelines that prioritize fairness, inclusivity, and the right to control one's data. For this, we can look to New Zealand, where Māori scholars and leaders have been working on decolonizing data governance through frameworks like Te Mana Raraunga[5] the Māori Data Sovereignty Network. This framework emphasizes the importance of indigenous peoples having control over how their data is collected, stored, and used, ensuring that it serves their community's needs and values. AI tools developed under this framework must comply with the ethical principles outlined by the Māori community, such as respect for collective ownership and the right to withdraw consent at any time.

In order to decolonize AI in Digital Humanities, frameworks like Te Mana Raraunga must be applied more broadly. Such frameworks allow marginalized communities to have control over their own data, ensuring that AI is used in ways that benefit them rather than reinforce existing power imbalances [13]; in addition they can guide how data from colonized regions is used in Digital Humanities projects, ensuring that the communities from which the data originates retain control over its use and interpretation.

4.4.2 Challenges and Solutions for Decolonial AI

While decolonizing AI in Digital Humanities is an urgent task, it is not without challenges. The technical design of AI systems, the availability of data, and the need for collaboration between technologists and humanists pose significant obstacles. However, the following solutions can guide the development of decolonial AI.

Community-Led Data Curation: Engaging marginalized communities in the process of data collection, curation, and use is critical. By ensuring that the communities from which data is sourced have a say in how that data is collected and used, AI systems can be designed to reflect their values and priorities rather than reinforcing colonial power dynamics.

Inclusive AI Design: AI systems should be designed with the inclusion of alternative epistemologies. This means not only integrating diverse data but also developing algorithms that can accommodate non-Western ways of organizing and interpreting

[5] Te Mana Raraunga: Māori Data Sovereignty Network, https://www.temanararaunga.maori.nz.

knowledge. This could involve designing NLP models that are trained on indigenous languages or developing machine learning techniques that respect oral traditions as valid forms of data.

Critical AI Literacy for Digital Humanists: Many digital humanists lack the technical knowledge to engage critically with AI systems. Building capacity in AI literacy among humanists can help bridge the gap between technological and ethical considerations. This would enable scholars to recognize when AI systems are reinforcing colonial perspectives and advocate for more inclusive and decolonial approaches to technology.

4.5 Conlusions

In this paper, we have explored the impact of AI-based techniques on Digital Humanities through a colonial perspective, highlighting how AI can both reinforce and challenge colonial power structures. AI's role in Digital Humanities has expanded the possibilities for analyzing large datasets, preserving endangered languages, and unlocking new interpretations of historical and cultural data. However, as we have seen, these advancements are not without significant ethical and epistemological concerns, particularly regarding the reproduction of colonial biases in AI systems.

One of the central arguments of this paper is that AI systems, particularly in Digital Humanities, are deeply influenced by the data on which they are trained and the algorithms they use. When these datasets and algorithms reflect colonial archives and Western knowledge systems, they risk reproducing the same power dynamics and epistemic exclusions that characterized colonialism. AI projects that rely heavily on colonial-era records, for instance, may prioritize the perspectives of colonizers while marginalizing or erasing the voices of the colonized.

At the same time, we have shown that AI offers powerful tools for decolonization, especially when designed with the explicit goal of including marginalized perspectives and alternative epistemologies. Projects like the Algonquian Linguistic Atlas and Māori Data Sovereignty initiatives demonstrate that AI can be used to preserve indigenous languages and knowledge systems, empowering communities to control their data and how it is represented.

To mitigate the risks of reinforcing colonial power structures, scholars, developers, and institutions working at the intersection of AI and Digital Humanities can adopt the following strategies:

- **Ethical AI Development:** Develop AI systems that are critically aware of the biases in their training data and seek to counterbalance these biases by incorporating diverse datasets and voices. This can include training NLP systems on indigenous languages or ensuring that visual recognition algorithms are not skewed toward Western art and cultural norms.

- **Community-Centric Approaches:** AI projects in Digital Humanities should engage directly with the communities whose data is being used, ensuring that these communities have control over how their knowledge is represented and used. This includes respecting indigenous data sovereignty, following ethical frameworks like Te Mana Raraunga, and fostering collaborations that benefit all parties.
- **Interdisciplinary Collaboration:** Bridging the gap between technologists and humanists is crucial for creating more equitable AI systems. Scholars in Digital Humanities should work alongside AI developers to ensure that ethical and epistemological concerns are embedded into the design of AI tools from the beginning. This interdisciplinary approach can help identify potential biases and blind spots in AI applications before they become entrenched.
- **Inclusive Epistemologies:** AI-driven Digital Humanities projects must move beyond Western frameworks of knowledge and embrace non-Western epistemologies. This involves recognizing the validity of oral histories, communal memory, and alternative understandings of history and culture. By integrating these perspectives into AI systems, Digital Humanities can help decolonize knowledge production and representation.

From a research point of view, further investigation is needed to develop robust methods for identifying and mitigating biases in AI systems, especially when applied to historical and cultural datasets, and make those mitigation strategies readily available as decolonial frameworks for AI. This work can be combined with an exploration of how alternative epistemologies can be integrated into AI systems and what ethical guidelines should govern these efforts. Finally, as AI technologies continue to evolve, long-term studies on their impact on marginalized communities, especially in relation to digital cultural heritage, are needed to assess whether AI is truly aiding decolonization efforts or simply repackaging old forms of control.

References

1. M. Aouragh, P. Chakravartty, Infrastructures of empire: Towards a critical geopolitics of media and information studies. Media, culture & society **38**(4), 559–575 (2016)
2. R. Benjamin, *Race After Technology: Abolitionist Tools for the New Jim Code*. (Polity, 2019)
3. N. Blomley, *Unsettling the city: Urban land and the politics of property*. (Routledge, 2004)
4. D. Chakrabarty, *Provincializing Europe: Postcolonial Thought and Historical Difference*. (Princeton University Press, 2000)
5. N. Couldry, U.A. Mejias, *The Costs of Connection: How Data Is Colonizing Human Life and Appropriating It for Capitalism*. (Stanford University Press, 2019)
6. N. Couldry, U.A. Mejias, Data colonialism: Rethinking big data's relation to the contemporary subject. Televis. New Media **20**(4), 336–349 (2019)
7. R. Cowie, E. Douglas-Cowie, K. Karpouzis, G. Caridakis, M. Wallace, S. Kollias, Recognition of emotional states in natural human-computer interaction. *Multimodal User Interfaces: From Signals to Interaction*, pp. 119–153 (2008)
8. R. Cowie, C. Pelachaud, P. Petta, *Emotion-oriented systems: the Humaine handbook*. (Springer, 2011)
9. D. Eltis, D. Richardson, *Atlas of the Transatlantic Slave Trade*. (Yale University Press, 2010)

10. M. Fricker, *Epistemic injustice: Power and the ethics of knowing*. (Oxford University Press, 2007)
11. M.-O. Junker, T. Stewart, A linguistic atlas for endangered languages: www. atlas-ling. ca, in *Edulearn11 Proceedings*, pages 3366–3376. IATED (2011)
12. K. Karpouzis, Plato's shadows in the digital cave: Controlling cultural bias in generative ai. Electronics **13**(8), 1457 (2024)
13. T. Kukutai, J. Taylor, *Indigenous data sovereignty: Toward an agenda*. (ANU Press, 2016)
14. W. Mignolo, *The darker side of western modernity: Global futures, decolonial options*. (Duke University Press, 2011)
15. S. Umoja Noble, *Algorithms of oppression: How search engines reinforce racism* (In Algorithms of oppression, New York University Press, 2018)
16. M. Pantic, G. Caridakis, E. André, J. Kim, K. Karpouzis, S. Kollias, Multimodal emotion recognition from low-level cues. *Emotion-oriented systems: The humaine handbook*, pp. 115–132 (2011)
17. C. Pinney, *Camera Indica: the social life of Indian photographs*. (University of Chicago Press, 1997)
18. A. Quijano, Coloniality of power, eurocentrism, and latin america. Nepantla: Views South **1**(3), 533–580 (2000)
19. R. Risam, *New Digital Worlds: Postcolonial Digital Humanities in Theory, Praxis, and Pedagogy*. (Northwestern University Press, 2018)
20. W. Edward, *Said* (Pantheon Books, Orientalism, 1978)
21. Y. Skarpelos, S. Messini, E. Roinioti, K. Karpouzis, S. Kaperonis, M.-G. Marazoti, Emotions during the pandemic's first wave: The case of greek tweets. Digital **4**(1), 126–151 (2024)
22. M. Walter, The voice of indigenous data: Beyond the markers of disadvantage. Griffith Rev. **60**, 256–263 (2018)

Chapter 5
Orbitorium: Designing and Curating a Digital Space Museum for Educational Engagement

Elpida Charitou

Abstract This chapter presents Orbitorium, a digital museum designed to engage audiences with the wonders of space exploration through immersive technologies. Built on the FrameVR platform, Orbitorium exemplifies the transformative potential of digital museums in educational outreach. The museum merges interactive exhibits, 3D models, and multimedia content to make complex astronomical concepts accessible to diverse audiences, from educators to students. Anchored in modern museology principles, the project explores the integration of interactivity, inclusivity, and human-computer interaction principles in curating digital spaces. The chapter discusses the design process, technical challenges, and opportunities for enhancing public engagement with science through virtual spaces. Orbitorium reflects broader trends in digital cultural management, showcasing how innovative platforms can democratize access to knowledge and foster deeper scientific literacy.

Keywords Digital museums · Space science education · Immersive technology · Interactive exhibits · Museology

5.1 Introduction

Orbitorium was developed as part of a university project at Panteion University of Social and Political Sciences, within the Department of Communication, Media, and Culture. The project was completed for the lab Contemporary Digital Applications in Culture under the Cultural Management track. This six-month endeavour provided a hands-on opportunity to explore the creation and curation of a digital museum, merging educational content with cutting-edge digital technologies to foster public engagement with space science.

This project not only allowed for the practical application of digital curation techniques but also explored the evolving role of digital technologies in cultural

E. Charitou (✉)
Panteion University of Social and Political Sciences, Athens, Greece
e-mail: charelpida7@gmail.com

K. Karpouzis and Y. Skarpelos (eds.), *Computational Methods for the Digital Humanities*, Learning and Analytics in Intelligent Systems 57,
https://doi.org/10.1007/978-3-032-05099-1_5

management. By merging theoretical insights with hands-on experience, Orbitorium showcases the potential of digital platforms in expanding cultural access and education.

Orbitorium also aligns with contemporary trends in digital culture, where museums increasingly embrace digital transformation to reach broader audiences. This project reflects the growing emphasis on interactivity, accessibility, and innovative educational methods in cultural institutions, providing a model for how digital technologies can transform traditional museum experiences. In the following sections, this chapter will explore the design and development of Orbitorium, focusing on the technical and conceptual frameworks that guided its creation. It will delve into the role of interactivity and user engagement, examining how human-computer interaction principles and immersive technologies were used to create an educational and interactive digital space. The chapter will also discuss the challenges encountered in curating vast amounts of scientific data for a digital platform, along with the opportunities that digital museums present in expanding accessibility and educational outreach. Finally, the conclusion will reflect on the lessons learned from this project and the future prospects for digital museums in fostering public engagement with science.

5.2 Contextualising Orbitorium

Orbitorium is a digital museum dedicated to space exploration, designed on the FrameVR platform, with the main goal of education. The museum's upper floor serves as an introductory space, offering visitors an accessible and engaging entry point to the vast topic of space. Rather than focusing solely on virtual space missions, the Orbitorium introduces fundamental astronomical concepts such as the solar system, meteorites, asteroids and the universe's structure. This introductory space is dedicated to making complex scientific ideas accessible to a wide audience, particularly students and educators, by incorporating multimedia elements such as 3D models, interactive exhibits, and audiovisual resources [6].

The design of the Orbitorium reflects the evolving role of museums, as digital learning environments. Rather than merely presenting static information, the platform engages visitors through interactive experiences that foster education and enjoyment, a concept often referred to as edutainment [3]. The digital nature of the Orbitorium allows it to expand beyond the limitations of physical museums, reaching a global audience that can access its educational content from anywhere, especially considering the growing trend for museums to adopt digital strategies to democratize access to knowledge [8].

The focus on education at the Orbitorium is based on the principles of contemporary museology, which recognises the museum not only as a static exhibition of objects but also as a place of learning and engagement [4]. By offering a personalized and immersive learning experience, the Orbitorium encourages visitors to interact with the material at their own pace, a key component of effective digital education [6].

The design and purpose of the museum highlight the potential of digital platforms to transform the educational process, making complex scientific concepts such as space exploration more accessible and engaging for different audiences.

5.2.1 The Educational Role of Museums

Museums have long played a crucial role in education, serving as spaces where the public can engage with history, art, science, and culture outside the formal classroom setting. Traditionally, museums have been seen as repositories of knowledge, but over the years, their role has evolved to become more dynamic and interactive. In modern museology, the emphasis is increasingly on creating learning environments that actively engage visitors, fostering curiosity, critical thinking, and deeper understanding, a shift that aligns with constructivist learning theories, which suggest that individuals learn best when they are active participants in their learning experience [9].

Digital museums, like Orbitorium, extend this educational mission by leveraging technology to reach broader audiences and offering interactive multimedia content that caters to diverse learning styles. The flexibility of digital platforms allows museums to present complex concepts, such as space exploration, in accessible ways. The design of the Orbitorium, with its audiovisual presentations and interactive exhibits, follows the edutainment trend, where educational content is combined with entertainment elements to keep the interest and engagement of the public [3].

In addition to making content more engaging, digital museums provide unprecedented access to educational resources, breaking down geographic and physical barriers. This is particularly important for reaching underserved or remote communities that may not have access to traditional museums. By offering interactive digital experiences, Orbitorium allows students, educators, and the public to explore scientific concepts from anywhere in the world, reflecting the broader educational goals of museums, which aim to democratize knowledge, foster lifelong learning, and inspire future generations of learners. As a result, digital museums like Orbitorium are crucial not only in preserving knowledge but also in promoting scientific literacy and public engagement with important contemporary issues.

5.2.2 The Role of Digital Museums in Modern Museology

Digital museums, such as the Orbitorium, are at the forefront of the transformation of museology. Generally, traditional museums are often limited by physical space and geographic accessibility, but digital platforms remove these barriers, allowing museums to reach broader and more diverse audiences globally [8]. In the case of Orbitorium, its virtual format enables visitors worldwide to engage with the educational content of space science without the need for physical attendance, thus enhancing access, participation and engagement [5].

The rise of new museology highlights the shift in museums' roles from static displays of artefacts to dynamic spaces of learning, interaction, and inclusion [4]. Digital museums are particularly well-suited to this shift because they offer interactive experiences that engage users in ways that traditional museums often cannot. Orbitorium, for instance, uses virtual reality, interactive spaces and 3D models to create an immersive educational experience, which allows visitors to explore space in personalised and interactive manner. Orbitorium, for instance, uses virtual reality and 3D models to create an immersive educational experience, which allows visitors to explore space in a personalised and interactive manner, while taking into account the growing influence of human-computer interaction (HCI) principles in museum design, which focus on creating engaging, user-friendly digital environments that encourage learning and discovery [2].

Furthermore, digital museums play a crucial role in addressing issues of accessibility and inclusivity. By leveraging digital platforms, museums can cater to audiences with diverse needs and backgrounds, marking a departure from the elitist origins of museums toward more democratic, open institutions. Orbitorium's design, for example, ensures that the museum is accessible to a wide range of users, including those who might face physical barriers to traditional museum visits [5].

5.2.3 *Significance of Space Museums*

Space museums hold a unique place in both education and public engagement with science. Whether physical or digital, they serve as critical platforms for disseminating knowledge about the universe, space exploration, and the technological advancements that have made such exploration possible. By presenting complex astronomical concepts accessible-wise, space museums play a key role in improving scientific literacy and fostering a broader understanding of humanity's place in the cosmos [1].

One of the main contributions of space museums is their ability to inspire curiosity and wonder about the universe while providing an understandable approach through interactive exhibits, 3D models, and educational programs for visitors who perceive space as a distant and abstract concept. Moreover, space museums contribute to preserving humanity's history of space exploration. Exhibiting materials related to missions like the Hubble Space Telescope and the James Webb Telescope ensures that the history of scientific achievements is documented and shared with future generations [1]. By acting as custodians of cultural and scientific heritage, space museums inspire future generations of scientists and innovators, embedding the values of exploration, discovery, and the pursuit of knowledge into the collective consciousness. Thus, the Orbitorium attempts to extend these benefits by making educational content about space exploration accessible without geographical and physical barriers.

Lastly, in addition to their educational mission, space museums also play a crucial role in connecting astronomy to broader cultural themes. Astronomy has been a fundamental part of human culture, shaping the way we view our existence in the universe. By presenting space science as part of this ongoing cultural narrative, space museums help bridge the gap between scientific knowledge and cultural understanding, making the exploration of the cosmos relevant to everyday life.

5.3 Research Question and Data Collection

This project is guided by the following research question: How can a digital museum effectively educate visitors about space exploration and scientific concepts through interactive and immersive digital experiences? This question aims to investigate the potential of digital platforms, such as FrameVR, to create engaging and educational environments. In particular, the research explores how a digital museum like Orbitorium can facilitate understanding complex astronomical concepts by employing interactive technologies, multimedia, and immersive design, while focusing on identifying strategies that make scientific content accessible to a wide range of audiences, from students to general visitors, while fostering curiosity and critical thinking. By addressing this question, the project seeks to highlight the educational potential of digital museums in promoting space science literacy and encouraging deeper engagement with scientific knowledge.

The data used to develop Orbitorium's exhibits was carefully curated to support the educational goals of the digital museum. The primary sources of information were derived from trusted and authoritative institutions, focusing on ensuring both scientific accuracy and accessibility. Much of the data was sourced directly from NASA's extensive digital archive, which provided high-quality educational content on various aspects of space exploration, including information about the solar system, galaxies, and cosmic phenomena. NASA's public domain resources, such as imagery and scientific data, were particularly valuable for building engaging and accurate visual exhibits.

In addition to NASA, other sources such as the BBC's digital platform were utilised, particularly for audiovisual material that could complement the static data sets. These external sources provided rich multimedia content, including documentary footage and explanatory videos about space missions, which were integrated into the museum's virtual space. The use of multimedia was central to making complex scientific concepts accessible to a broad audience, especially for younger visitors and those unfamiliar with the subject matter.

The choice of data was based on the need to present the information in an interactive and multilevel way. This involved translating complex scientific data into user-friendly formats such as 3D models, visual infographics and interactive displays, designed to promote active learning. By carefully selecting content that balanced scientific depth with accessibility, with Orbitorium I attempted to create a digital space that is educational, engaging and easy to navigate.

5.4 The Design and Technical Process

5.4.1 Designing Orbitorium

The design of Orbitorium, a digital space museum created on the FrameVR platform, was a comprehensive process that balanced educational goals with technological possibilities. The central objective of the project was to make the complexities of space science accessible and engaging to a wide range of audiences. To achieve this, the design process focused on creating an interactive and immersive digital environment where visitors could explore astronomical concepts in a structured and visually appealing way (Fig. 5.1).

The digital museum consists of two main areas, one dedicated to the NASA space missions "Hubble Space Telescope" and "James Webb Telescope" which is under construction, and the second area is the one that will be analysed in this chapter. Upon entering the Orbitorium, visitors are immediately immersed in an amphitheatrical space designed to introduce them to the general concepts of space and the universe. This introductory area features a video that provides broad information about galaxies, setting the stage for more detailed exhibits to follow. The spatial arrangement of the museum is carefully curated to facilitate a progressive understanding of astronomical concepts. The central area of the exhibition is structured thematically, beginning with a general framework of the universe's elements–such as stars, planets, and galaxies–before moving towards more specific exhibits. The exhibition space includes a variety of educational resources, such as tables of planetary

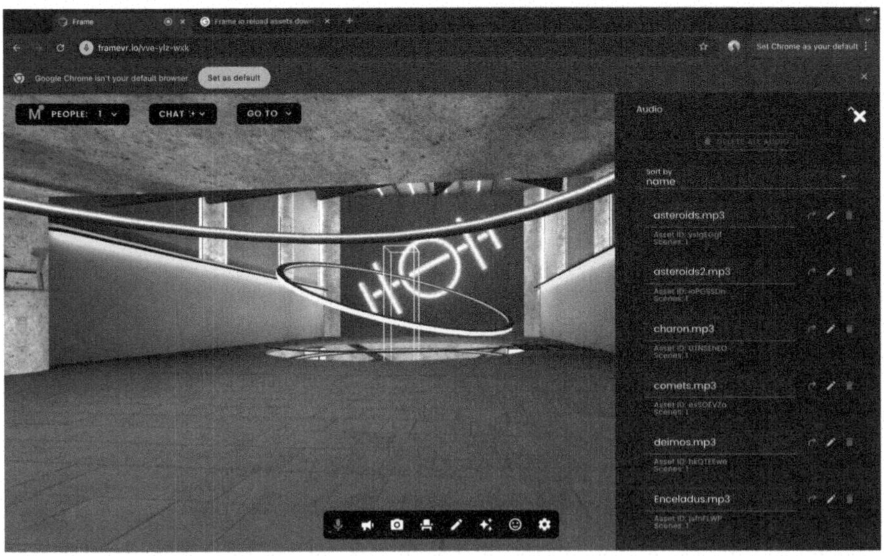

Fig. 5.1 A screen from the Frame VR environment

data, 3D models of satellites, and high-quality photographic materials that provide visual representations of key astronomical phenomena. The systematic use of audio material accompanying each exhibit was chosen to enhance the visitor's experience, offering specific, detailed information that complements the visual content. Although the exhibits are housed within a single, unified space, they are arranged thematically in a manner that encourages logical progression through the museum. This thoughtful organisation allows visitors to easily orient themselves and understand how the exhibits are interrelated, creating a cohesive narrative that guides them through the museum.

5.4.2 The Conceptual Framework of the Museum

The conceptual foundation of Orbitorium was built on principles of new museology, which emphasise user-centred experiences, interactivity, and inclusivity [7]. A key element of this approach was the decision to divide the museum into distinct thematic areas, each offering a different entry point into space science. The upper floor of the museum, designed as an introductory space, focuses on basic astronomical concepts such as the solar system, stars, and galaxies. This space was specifically designed for educational purposes, targeting both students and general visitors who may be unfamiliar with the subject matter. The goal was to introduce these concepts in a simplified yet accurate manner, ensuring that visitors could engage with the content regardless of their prior knowledge [6].

A major consideration in the design of Orbitorium was the role of storytelling in guiding visitors through the exhibits. The layout of the museum was crafted to take users on a structured journey, beginning with general information about the universe and gradually delving into more specific topics like planetary data and astronomical phenomena. This narrative arc was essential for maintaining coherence and providing a logical progression of knowledge, helping visitors make connections between different concepts [8].

5.4.3 Technical Development

In terms of the use of technology and interactivity, the design of the Orbitorium made extensive use of interactive technologies to create a more engaging and immersive experience for visitors. FrameVR, the platform used for the museum, offers extensive capabilities that allow for the use of virtual reality (VR) and 3D modelling. These features allowed the museum to incorporate interactive 3D exhibits, such as models of planets and satellites, which visitors could explore in detail. This interactive element was a critical part of the design, as it transformed the museum from a static collection of information into an engaging, hands-on learning environment [3]. The immersive potential of FrameVR's virtual reality features was also utilised to enhance visitor engagement. The museum was designed to be compatible with VR headsets, which allow visitors to navigate the space as though they were physically present in a traditional museum. This feature added a layer of interactivity and immersion that is often lacking in digital educational platforms. Visitors are not only passive recipients of information but active participants in their learning experience, able to explore exhibits at their own pace and interact with various elements according to their interests [6].

One of the design challenges was ensuring that the digital space was intuitive and easy to navigate. To address this, the layout of Orbitorium was designed to be both thematic and user-friendly. Upon entering the museum, visitors find themselves in a central amphitheatre-like space where they are introduced to general information about the universe through videos and multimedia presentations. From this central space, visitors can access different areas of the museum, each dedicated to specific topics like planets, galaxies, or space missions [8].

The user interface was another critical aspect of the design. Since digital museums lack the physical cues that guide visitors in traditional spaces, ensuring intuitive navigation was essential. FrameVR's platform allowed for the integration of clear visual markers and instructions, making it easy for visitors to understand where they are within the museum and how to interact with the exhibits. This user-centred approach to design, rooted in human-computer interaction principles, ensured that the digital space was accessible and engaging for all users [2].

5.4.4 Educational Objectives of the Museum

The educational goals of Orbitorium were central to every design decision. The museum's content was curated to provide a balance between entertainment and education, commonly referred to as edutainment [3]. Translating complex scientific concepts into easily understandable forms, and using a variety of media to accommodate different learning styles was at the core of the museum's design. For example, visitors have the opportunity to engage with 3D models for visual learning, watch videos for auditory learning and participate in 'discussion tables' where they can interact with other visitors.

At this point, I would like to focus on the presence of the interactive table, designed as a virtual space where visitors can exchange ideas, collaborate and engage in meaningful discussions. This space, designed primarily for educational purposes, facilitates a dynamic learning environment, allowing teachers to conduct lessons within a fully immersive digital context. The interactive table promotes collaborative learning, a key element of contemporary educational pedagogy, by allowing students to collaborate in real-time, explore scientific concepts and enhance critical thinking through discussion and interaction [6]. This feature was created based on current trends in digital learning environments, where digital museums are not just seen as repositories of knowledge but as active learning spaces that support formal education. By providing a platform where visitors, including teachers and students, can interact both with the exhibits and with each other, the Orbitorium supports a constructivist approach to learning, where knowledge is built through social interaction and experience [8]. It is, after all, important in today's digital age to promote an environment where education is not limited to passive learning, but is instead enhanced by active engagement and collaboration, making Orbitorium a flexible educational resource.

In conclusion, the design of Orbitorium was a multifaceted process that combined educational, technological, and user-centred considerations. By leveraging the capabilities of digital platforms and interactive technologies, the museum successfully created an immersive learning environment that aligns with the principles of modern museology. The careful balance of content, interactivity, and user experience ensures that Orbitorium not only educates its visitors but also inspires further exploration and curiosity about space science.

5.5 Discussion

5.5.1 Challenges in Digital Curation

Digital curation for a project like Orbitorium presents unique challenges, particularly in terms of ensuring that the content is both accessible and engaging while maintaining scientific accuracy. One of the key issues in digital curation is the management and organisation of vast amounts of data. For Orbitorium, sourcing and curating relevant, accurate, and high-quality materials from institutions like NASA and the BBC was essential but also posed significant challenges.

Another challenge in curating Orbitorium was the need to translate a vast amount of complex information manually from NASA's digital resources. Translating this content posed logistical and practical challenges, as it required sifting through NASA's digital archives and selecting information that was relevant and engaging for the Orbitorium audience. While NASA provided a wealth of scientific data, exhibits, and multimedia on space exploration, this information needed to be not only curated but also translated into accessible, user-friendly formats that could engage visitors of varying backgrounds and levels of expertise. The translation process was extensive and required careful attention to ensure that the scientific integrity of the material was preserved, while also simplifying it enough for broader public understanding.

Additionally, digital curation must consider the usability and navigation of the museum. Unlike physical spaces, digital museums lack physical cues that guide visitors, making it essential to curate the experience with clear visual markers and intuitive navigation paths. In Orbitorium, ensuring that visitors could easily locate and interact with exhibits without becoming overwhelmed by the volume of content was a primary challenge. The platform's user interface needed to be designed with human-computer interaction principles in mind, ensuring accessibility for users with varying levels of digital literacy.

5.5.2 Interactivity and User Engagement

The concept of interactivity is central to the Orbitorium's design, transforming the museum from a passive collection of knowledge into a dynamic space for active exploration and learning. Interactivity in digital museums is particularly effective due to its ability to engage visitors through human-computer interaction principles and immersive technologies, allowing for a more personalised and engaging experience.

Human-Computer Interaction plays a critical role in making digital museums such as the Orbitorium user-friendly and engaging. In a digital space where physical signs and traditional methods of guiding visitors are absent, intuitive navigation becomes essential. The Orbitorium addresses this challenge by implementing clear visual markers, interactive touchpoints and distinct thematic modules that help visitors navigate the museum and interact with the exhibits effortlessly [2]. Incorporating HCI principles ensures that visitors can engage with the museum regardless of their digital knowledge, making the space accessible to a wide audience. These features are particularly important in educational contexts, where seamless navigation and engagement are essential to promote learning.

The use of interactive technologies such as 3D models and multimedia presentations in the Orbitorium allows visitors to interact with complex scientific concepts engagingly. Visitors are encouraged to explore satellite models, audio data, 3D representations of stars and planets and interact with each other at a digital meeting point, hence creating a learning experience that is not only educational but also participatory. Therefore, by basing its design on HCI principles, Orbitorium ensures that the user journey is both informative and interactive, allowing for a deeper engagement with the subject matter [6].

5.5.3 Opportunities and Challenges in Digital Museums

Digital museums, such as Orbitorium, present significant opportunities for expanding the reach and impact of museums in the digital age. By leveraging technology, these museums can break down traditional barriers of geography, accessibility, and engagement, offering rich, interactive learning experiences to global audiences. However, the implementation of digital museums also brings a unique set of challenges that must be addressed to ensure they remain effective and inclusive educational tools.

One of the most significant opportunities presented by digital museums is their ability to democratise access to knowledge. Traditional museums are often geographically limited, meaning that individuals who cannot physically visit these spaces are excluded from their resources. Digital museums eliminate this barrier, allowing users from any part of the world to engage with educational content at their convenience. Orbitorium, for example, makes space exploration content accessible to global audiences, fostering scientific curiosity and understanding across a wide demographic [8].

Another key opportunity is the ability to integrate multimedia and interactive technologies. Digital museums have the flexibility to incorporate a range of media formats, including 3D models, virtual reality, videos, and interactive exhibits, making learning more engaging and adaptable to various learning styles [6]. These technologies enhance the educational impact of digital museums by enabling visitors to interact with content in ways that are not possible in physical spaces. Immersive experiences, such as those provided by virtual reality in Orbitorium, allow visitors to feel as if they are physically exploring space, which deepens their connection to the subject matter and enhances the learning experience [8].

Digital museums also provide opportunities for ongoing content updates and improvements. Unlike physical museums, where exhibits may remain static for years, digital museums can easily update their content, incorporate new scientific discoveries, and respond to user feedback. This dynamic approach ensures that the information remains current and relevant, allowing museums to continuously engage with their audiences [7].

Despite these opportunities, digital museums also face several challenges. One of the most pressing issues is the digital divide, which refers to the disparity in access to digital technologies across different populations. While digital museums can theoretically reach global audiences, they are only accessible to those who have reliable internet access and the necessary technological tools, such as computers or VR headsets [3]. This limits the ability of digital museums to truly democratise access to knowledge, as underserved populations may still face barriers to engagement.

Another challenge lies in maintaining user engagement. In traditional museums, visitors can physically interact with objects and navigate through exhibits, creating a tangible experience. Digital museums, however, must work harder to sustain user interest and engagement in a virtual environment. Without the physical presence of artefacts or the atmosphere of a museum space, digital exhibits must rely heavily on interactive features, storytelling, and multimedia to keep visitors engaged [3]. Creating a balance between informative content and engaging design is crucial but can be difficult to achieve, especially given the vast amount of information that must be presented in a cohesive and accessible way.

Digital curation also poses challenges in terms of content management. Curating vast amounts of data in a digital space, while ensuring that it is well-organised, accurate, and accessible, can be a complex task. For Orbitorium, the manual translation of NASA's extensive scientific data into user-friendly formats was a major challenge [6]. This issue extends to ensuring that the digital space is intuitive and easy to navigate, particularly when digital museums do not benefit from the physical layout cues of traditional museums. Thoughtful user interface design and clear visual markers are necessary to prevent visitors from becoming disoriented or overwhelmed by the digital content [8].

In conclusion, while digital museums like Orbitorium offer significant opportunities to expand access to knowledge and engage visitors through innovative technologies, they also face challenges related to digital access, user engagement, and content management. Addressing these challenges will be key to ensuring that digital museums can fulfil their potential as inclusive, dynamic spaces for learning and cultural engagement.

5.6 Conlusions

The creation of Orbitorium was a comprehensive project designed for the lab Contemporary Digital Applications in Culture, taught by Kostas Karpouzis. Over the course of six months, the museum was meticulously developed, blending educational goals with innovative digital technologies. The process offered a valuable opportunity to explore the practical aspects of curating a digital museum, from sourcing and organising complex scientific data to building an interactive and immersive space for users. Orbitorium stands as a testament to the prospects of digital museums and the transformative potential they hold for engaging audiences, making science more accessible, and fostering a deeper understanding of space exploration.

Orbitorium was designed to create a user-friendly, interactive environment that could serve as a powerful educational tool. The digital platform utilised multimedia elements, such as 3D models, virtual reality, and audiovisual content, to bring complex astronomical concepts to life for a wide audience. By focusing on interactivity, personalisation, and accessibility, Orbitorium exemplifies the potential of digital museums to democratise access to scientific knowledge and inspire curiosity in diverse populations. Throughout its development, the project addressed challenges typical to digital curation, such as translating vast amounts of scientific data and ensuring that the museum remained intuitive and engaging for users.

The development of the Orbitorium provided invaluable insights into the practical challenges and opportunities of digital curation and museum design. One of the key lessons learned was the importance of balancing educational content with user engagement. The project demonstrated that making complex scientific information accessible to a broad audience requires careful translation and presentation, especially in digital environments. Additionally, the project highlighted the significance of intuitive user interface design, ensuring that visitors of varying digital literacies could navigate the space with ease. The process of manually translating and curating vast amounts of scientific data also underscored the need for meticulous organisation and attention to detail in digital museum projects. Most importantly, this project reinforced the potential of digital museums to foster curiosity and enhance public engagement with science through innovative and interactive technologies, providing a valuable model for future digital cultural initiatives.

Looking ahead, the future of digital museums like Orbitorium is bright. As technology continues to advance, there will be greater opportunities for enhancing user engagement through even more immersive experiences. Expanding the museum's content, incorporating real-time data updates, and integrating additional collaborative tools for visitors will further strengthen its educational impact. Furthermore, the lessons learned from this project provide a foundation for future digital curation efforts, highlighting both the challenges and rewards of creating digital spaces that are accessible, educational, and engaging. Orbitorium stands as a forward-looking model for how digital museums can evolve to meet the changing needs of both educational institutions and the general public.

References

1. A. Acker, Planetariums, theatres of the universe. Proc. Int. Astron. Union **5**(S260), 465–474 (2009)
2. E. Hornecker, L. Ciolfi, *Human-Computer Interactions in Museums* (Springer, Cham, Switzerland, 2019)
3. A.-X. Koutoura, Edutainment in museums: a tool for disseminating knowledge. Museum Manage. Curatorship **29**(1), 66–82 (2014)
4. S. Zardini Lacedelli, *The platform-museum: conceptual revolutions and practical implications* (Museums of the Future, Museological Review, 2018)
5. E. Lisney, J. Bowen, K. Hearn, M. Zedda, Museums and technology: Being inclusive helps accessibility for all. Curator: Museum J. **56**(3), 353–361 (2013)
6. N. Mamur, V. Özsoy, I. Karagöz, Digital learning experience in museums: Cultural readings in a virtual environment. Int. J. Contemporary Educ. Res. **7**(2), 335–350 (2020)
7. M.D. Osterman, Museums of the future: embracing digital strategies, technology and accessibility. Museological Rev. Museums Fut. **22**, 10–17 (2018)
8. C. Wilson-Barnao, *Digital access and museums as platforms*. Routledge (2021)
9. K.A. Yellis, Museum education, in *The Museum: A Reference Guide*, ed. by M.S. Shapiro. (Greenwood Press, Westport, CT, 1990)

Chapter 6
Eye Tracking in Immersive Environments for the Assessment of Key Drivers Affecting Food Consumption Habits

Nikolaos D. Tantaroudas, Kostantinos Koukoudis, Georgios Tsimiklis, Ioannis Karaseitanidis, Angelos Amditis, and Tina Katika

Abstract The work presented in this chapter contributes to the development of a systematic method for analyzing human preferences and decision-making by utilizing eye-tracking technology within extended reality (XR) environments. Participants are exposed to diverse food consumption scenarios using modern video see-through headsets within extended reality (XR) environments. Initial validations of the methodology are performed within the VR environment, where comprehensive eye-tracking data is captured and compared against responses provided from questionnaires, enhancing the understanding of participants' food choices. Demonstrations in an MR environment environment enhance data storytelling by integrating key factors influencing food consumption habits. This approach reveals underlying preferences and biases often missed by traditional surveys, providing valuable insights into consumer behavior. Preliminary findings suggest that eye-tracking data reveals a link between participants' gaze and their food choices, influenced by their backgrounds and dietary habits. Different dietary groups–carnivores, vegetarians, and omnivores–show distinct patterns of engagement with food scenarios, demonstrating how mixed reality (MR) technology can be used to customize educational and marketing approaches based on individual preferences.

Keywords Virtual reality · Mixed reality · Extended reality · Eye tracking · Food preferences assessment

N. D. Tantaroudas (✉) · K. Koukoudis · G. Tsimiklis · I. Karaseitanidis · A. Amditis · T. Katika
Institute of Communication and Computer Systems, Zografou, Greece
e-mail: nikolaos.tantaroudas@iccs.gr
URL: https://www.iccs.gr/el/

K. Koukoudis
e-mail: k.koukoudis@iccs.gr

G. Tsimiklis
e-mail: georgios.tsimiklis@iccs.gr

A. Amditis
e-mail: gkaraa.amditis@iccs.gr

T. Katika
e-mail: tina.katika@iccs.gr

K. Karpouzis and Y. Skarpelos (eds.), *Computational Methods for the Digital Humanities*, Learning and Analytics in Intelligent Systems 57,
https://doi.org/10.1007/978-3-032-05099-1_6

6.1 Introduction and Related Work

6.1.1 Eye-Tracking Extended Reality Technology in Market Research and Consumer Analysis

The emergence of XR eye-tracking technology has become a popular area of study and commercial interest with diverse applications spanning across various fields [1–5] alongside significant use in marketing and consumer behavior analysis [6]. In market research and consumer behavior analysis, XR eye-tracking enables researchers and industry players to analyze how users interact with XR environments, providing valuable data on visual attention, engagement, and decision-making processes. This innovative technology offers a deeper understanding of consumer behaviour, allowing for the development of more effective marketing strategies and improved product designs [7].

Eye-tracking has been extensively used to study consumer behavior at points of sale, revealing how visual information is processed during purchasing decisions, highlighting the impact of packaging design on decision-making phases [8]. Clement et al. [9] assessed the effectiveness of various packaging features including shape, size, color, and font, as well as the positioning of nutrition labels and the influence of misleading elements, demonstrating that packaging and advertising designs significantly influence consumer behavior and preferences. Moreover, eye-tracking technology has played a significant role in understanding the importance of product pricing and placement. Existing studies has shown that the strategic placement of products, alongside relevant information and pricing, affects purchase intentions [10, 11]. In these scenarios, eye-tracking revealed that the strategic location of products significantly impacts consumer choices, essential for managing retail shelves and in-store promotions. Additionally, studies on peripheral vision's role in viewing retail displays indicate the complex ways consumers engage with their visual environment during shopping [12].

The use of eye-tracking has revealed visual cues that affect consumers' behaviour at point of sales and even goes further into the analysis of emotions. Research in this field has focused into how emotional states influence decision-making processes while shopping. For example, Gunaratne et al. [13] studied how emotions impact the use of decision rules such as satisfaction or maximization when consumers select food items. The study found that positive moods do not change decision strategies but result in detailed and slow product scrutiny without altering consumers' choice tactics with items. Moreover, eye-tracking has provided insights into how different shopping objectives and mental heuristics affect visual attention during purchasing decisions, highlighting its critical role in understanding and influencing consumer behavior [14].

Eye-tracking technology aside from demonstrating its worth in retail environments, has proved an effective component of broader marketing research which includes evaluating trade fair booths, TV commercial and printed advertising [15]. In television advertising, Smink et al. [16] applied eye-tracking technology to study the brand placement and viewer engagement, investigating how brand familiarity affects message reception and brand recall. In printed advertising, eye-tracking has been used to examine how different typeface designs influence viewer attention and attitudes toward advertisements and brands [17]. Additionally, Hutton et al. [18] researched how gaze cues from models in ads direct viewers' focus to specific ad areas, contributing to our understanding of visual elements in advertising effectiveness.

Building on these insights, eye-tracking has been extensively applied to the study of modern online advertising, revealing important details about how consumers interact with digital advertisements. Research has focused on how the alignment of ad banner content with webpage content affects user engagement and memory recall, focusing on metrics such as the number of fixations, duration of fixations, and the timing of the first fixation [19]. Research in the area of online advertising has also examined "banner blindness" [20], where tools such as eye-trackers determine whether users notice ads on platforms such as Facebook or on e-commerce websites compared to other content, such as friend recommendations or familiar visual elements. In addition, the phenomenon of "text ad blindness" has been also analyzed [21, 22], investigating how users search for and notice text ads on websites. This research has helped understand the impact of visual effects such as animations on user attention and how these effects may influence cognitive processing.

Although eye-tracking is a versatile method in market research, useful for evaluating product packaging, website design, and effectiveness of various marketing communications, it is often integrated with other methods, including quantitative surveys and qualitative techniques like focus groups in order to enhance the depth of consumer behavior analysis [23]. This approach is also commonly used together with other neuromarketing techniques to provide a more comprehensive understanding of consumer interactions with advertisements and content. Eye-tracking technology, when integrated with VR, MR, and XR, provides detailed insights into user attention and behavior, offering a dynamic tool for market research. This technology facilitates an immersive experience where researchers can analyze eye movement patterns, gaze fixation, and pupil dilation to infer user preferences and decision-making processes.

6.1.2 Food Habits Assessment in Extended Reality Environments

Traditional methods such as questionnaires, interviews, and food diaries, have long been used to assess food habits, offering valuable insights into dietary patterns, preferences, and nutritional knowledge. However, these approaches often struggle with biases due to self-reporting, which can lead to inaccuracies regarding actual food intake and preferences [24]. Moreover, while useful for large-scale studies, questionnaires frequently capture only minimal information about dietary intake, failing to uncover the key reason why individuals make specific food choices [25]. Despite these limitations, past studies utilizing questionnaires have effectively demonstrated the correlation between food habits and various health outcomes, such as obesity and physical activity levels, emphasizing the significance of self-reported data in revealing the complexities of individual dietary behaviours [26].

Recent advances in Extended Reality (i.e. Virtual, Augmented and Mixed Reality (AR/VR/MR)) technologies now allow for more controlled and realistic settings to study these choices more accurately [27]. Immersive experiences enable participants to engage with virtual food items, providing comprehensive data on visual attention, selection methods, and consumption habits. This method facilitates the observation of individual behaviors in a simulated real-world setting, yielding insights into the decision-making processes that guide specific food choices. The integration of eye-tracking technologies with XR has significantly improved research capabilities by providing objective measures of an individual's visual attention, including gaze patterns, fixation duration, and pupil dilation [28, 29]. These metrics deliver direct observations of cognitive and emotional reactions to various food stimuli, which typically occurs subconsciously and are difficult to measure accurately through traditional self-reporting methods.

Initial investigations have utilized VR and eye-tracking to assess food-related behaviours, suggesting these technologies' potential in promoting behavioural shifts towards sustainable eating habits [28, 29]. Ruppenthal et al. [30] conducted a systematic review of eye-tracking research focused on sustainable food consumption, offering a detailed analysis of the interactions between visual attention and consumer behavior in food selection scenarios. Additionally, VR has been applied to examine how immersive settings affect food cravings, indicating a potential to influence eating behaviours by altering food-related sensory cues [31]. Allman-Farinelli et al. [32] developed a VR food court to explore meal choices among young people, showcasing VR's effectiveness in simulating real-world food environments for evaluating dietary behaviours. Further studies have assessed the visual quality of food in VR, indicating that realistic virtual representations can influence individual food choices [33]. Research efforts have also extended into promoting healthy eating habits among younger demographics using VR, emphasizing the technology's potential in shaping early dietary habits [34].

Despite recent progress, the use of MR technologies, in directly assessing food habits has not been thoroughly explored. Initial findings from studies using VR to analyze behavioural aspects of food choice show promise [35, 36], but comprehensive evaluations of food habits using MR methods are still limited [37]. In the present study, the MR technology introduced aspires to extend the VR capabilities by merging physical and virtual elements. Traditional questionnaire methods are combined with advanced MR technologies and eye-tracking features. Initial validations of the methodology are performed within a VR environment, where comprehensive eye-tracking data is captured and compared against responses from questionnaires, enhancing the understanding of participants' food choices. Furthermore, demonstrations are carried out in a MR environment and key factors affecting food choices are delivered through immersive data storytelling visuals. This integration of key factors influencing food consumption habits reveals underlying preferences and biases often missed by traditional survey techniques, providing invaluable insights into consumer behavior.

By focusing on essential eye-tracking metrics such as gaze direction, duration, and frequency, the study identifies which food options attract individual attention. This approach uses heatmaps to visualize attention points, enriching the understanding of dietary behaviors by identifying crucial factors that promote sustainable and healthy eating habits.

6.2 Methodology

6.2.1 Background

When analyzing eye-tracking data in VR environments, it is essential to consider specific metrics like gaze direction, duration, and the number of fixations. Gaze direction provides valuable insights into participants' visual focus and attention distribution [38]. Meanwhile, the duration of gaze and the frequency of fixations offer information about the extent and frequency of participants' engagement with specific elements in the virtual space. Previous studies have emphasized the significance of these metrics in understanding visual interactions and decision-making processes [39]. Another crucial aspect of the analysis is determining the threshold for gaze duration. Extremely brief glances, such as those lasting only 0.1 seconds, may not indicate meaningful engagement or interest. In the current study we built on top of past work by setting a threshold to filter out random fixations and concentrate on conscious visual attention [40]. Through this approach, we can differentiate between random glances and important attention points and gain a better understanding of participant interactions within the XR.

6.2.2 Description and Experimental Setup

The current work distinctively leverages XR hardware and the Movement SDK API, alongside other agnostic frameworks, to perform an initial validation in a VR setting and facilitate a seamless demonstration into engaging Mixed Reality settings (MR). The method involves collecting precise measurements of participants' gaze behavior while interacting with virtual food stimuli, with the goal of understanding how visual engagement affects food choices. This approach is facilitated by collecting data on gaze location, duration, and point of fixation, which can be analyzed and plotted through heatmaps to indicate areas of visual interest.

The study is divided into consecutive phases, starting with comprehensive questionnaires introduced to collect data on participant demographics, lifestyle, dietary preferences, and health issues. Following this step, participants representing spe-

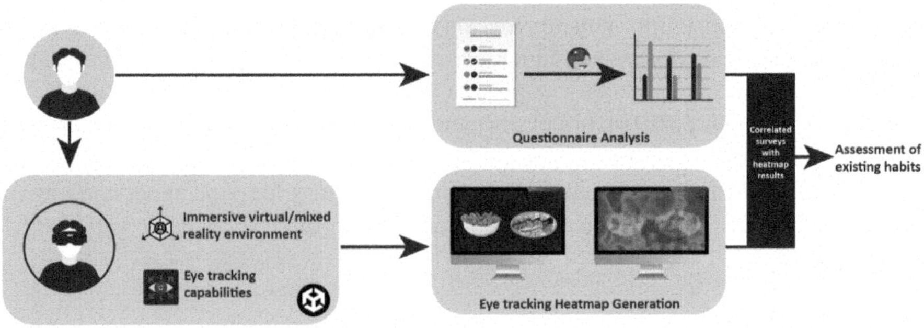

Fig. 6.1 Extended reality-based behavioral analysis flowchart: Identifying dietary choices through eye-tracking, HeatMaps and data analysis

cific food habit groups are exposed in an immersive VR experience where their interactions with different food images are monitored using eye-tracking technology delivered by the headsets. This setup allows for gathering quantitative insights into participants' visual attention patterns for different food choices and correlating this with self-reported preferences.

Eye-tracking metrics such as fixation duration, frequency, and fixation point are collected and analyzed. The data are further improved by converting quaternion rotation data into Euler coordinates to allow precise analysis of gaze direction and focus. The use of quaternions ensures accuracy in representing 3D rotational motion, which is crucial for identifying specific areas in a VR environment that capture the participant's attention. The ultimate goal is to reveal and understand the psychological processes behind unconscious preferences and food choice decisions. The experimental flowchart detailing the distinct phases of the methodology is given in Fig. 6.1. The flowchart provides the study's comprehensive and innovative approach to exploring the complex dynamics between visual attention and dietary decisions. This particular setup was chosen to facilitate the implementation of a proof-of-concept targeting different food choices, designed to assess the food preferences of the participants in an unbiased and inclusive manner.

6.2.3 Definition of Scenarios

The definition of the scenarios in an immersive setup involves several steps, including the design of the virtual environment, the selection and placement of the options as well as the selection of the most appropriate environment to perform the assessment. This present study develops a functional prototype capable of drawing initial conclusions on the factors influencing food choices and supporting hardware and software agnostic frameworks that are transferable to other settings. In the initial phase, photos of different food items were placed against a wall in a virtual room and eye tracking data were collected. This foundational phase is necessary to ensure that the current flexible and transferable methodology and technology is validated and tested before shifting into MR scenarios that are more complex and challenging at once. Following the previous validation, a more complex scenario is designed to assess the unbiased food preferences of participants by testing various food options in a MR environment. This includes creating the MR space and selecting appropriate food items to be displayed on a table, complemented by immersive 3D visuals for data storytelling. This setup is intended to facilitate eye-tracking studies within the MR environment, enhancing the assessment of how participants interact with different dietary choices and key factors affecting food habits.

6.2.4 Eye-Tracking Data Collection and Analysis

The Unity development environment includes the OVR Eye Gaze component that Oculus has developed into a game object that represents the user's eyes. This component is responsible for tracking either the left or right eye with the help of a camera located on the Quest headset and detecting position and rotation data. As a result, an object from which we retrieve OVR Eye Gaze data Fig. 6.2 shows us the true movement and orientation of the human eye.

Following the generation of the eye rotation data, a custom script is implemented to direct a ray from the focal point of an animated eye (game object) towards a specific direction, at a given frequency (e.g., 50 Hz). The resulting ray-casting outcomes are then analyzed to identify interactions between objects within the virtual environment. Once the ray intercepts the object that is considered of interest (e.g., an image of a food item) and remains focused for more than a certain period of time, such as 0.15 seconds, the timer will begin recording the event. Moreover, in order to keep track of where a ray hits an object, regardless of the type, a collection of 3D points is preserved. The coordinates (x, y, z), the timestamp of the collision, the user clicks on the immersive content, and the name of the object involved are entered into a list. By collecting such data, an analysis after processing is carried out to study user gaze behavior in a mixed reality world down to details.

6.2.5 Analysis of Experimental Visual Attention and Self-Reporting Data

The analysis of the experiment focuses on visual attention data, which is collected from eye-tracking metrics in XR. This data is then matched with self-reported questionnaires to identify the individuals' trends and preferences related to their food choices. In addition, the analysis produces three-dimensional heatmaps to capture and depict the data gathered from eye tracking. These heatmaps are able to effectively emphasize the parts of the XR environment that receive the most attention, thus helping to highlight where people's main interest lies and where their primary

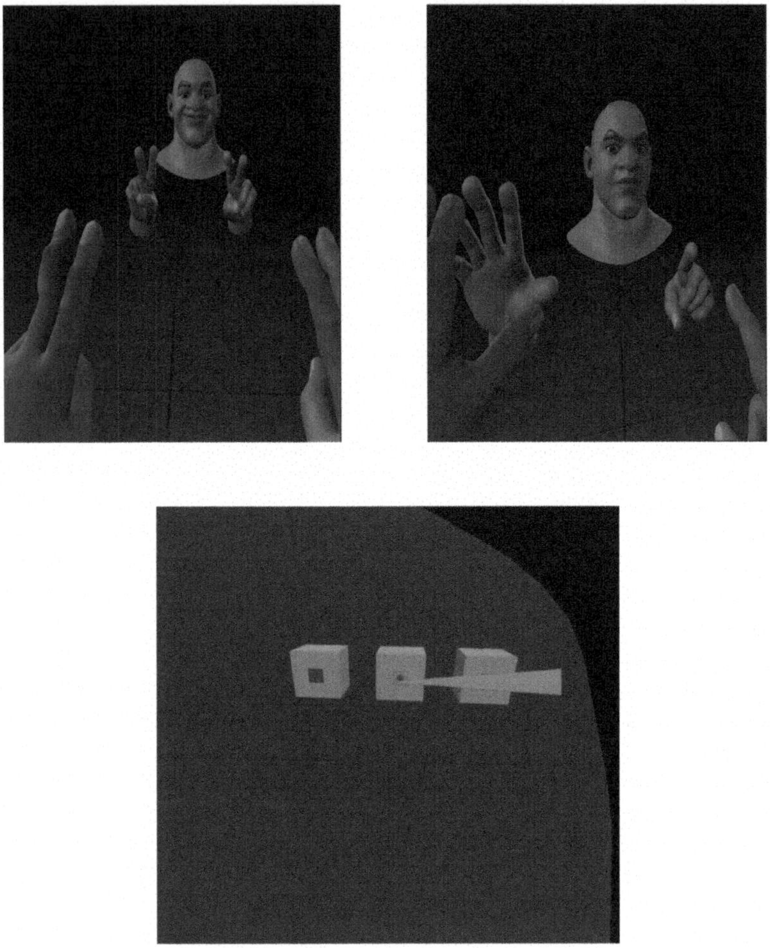

Fig. 6.2 Sensors setup and Avatar testing and calibration for both scenarios

engagement originates. Also, by overlaying these points of attention on a background of food items, researchers can determine which ones attract the highest visual focus, thus inferring higher levels of interest or preference.

On the basis of this visual attention information and the self-reports from questionnaires, it can be determined that it is possible to study food habits in a more subjective manner, relying on quantitative rather than qualitative metrics. The questionnaires collect structured and detailed information, including diet restrictions, health, ethical issues, cultural impact, and personal preferences. By comparing the explicit preferences and inclinations in the questionnaires to the implicit interest in the MR environment, we are able to identify the drivers of food selection behaviors. This analysis does not focus on the specific food choices used in the experiment. Instead, the emphasis is on the methodology itself, which is designed to be broadly applicable across various dietary choices and settings. Therefore, six food choice scenarios identified were selected based on informal interviews with representatives of each group.

6.3 Results

6.3.1 Assessment of Participants

A thorough data gathering effort was conducted, involving 115 participants who were recruited for the study. The process began with a detailed questionnaire designed to gather data on participants' demographics, lifestyle choices, dietary preferences, and health-related factors. Table 6.1 offers a concise summary of the food choices made by various demographic groups among the respondents. The evaluation using traditional questionnaires revealed a wide variety of food preferences, which underscores the complexity of behavioral patterns related to food choices and emphasizes the importance of including a broad spectrum of dietary practices in such studies.

Significant variations are evident among carnivores, vegetarians, and vegans regarding gender distribution, body mass index (BMI) categories, health issues, and considerations of environmental impact. Table 6.2 indicates a higher proportion of males in the carnivorous diet group and a lower proportion in the vegetarian and

Table 6.1 Questionnaire results summary

Category	Results	Category	Results	Category	Results
Number of participants		Gender		Education level	
	115	Male	43	High-school	5
Age groups		Female	70	College	10
18–24	8	Non-binary	0	Bachelor	28
25–34	25	Prefer to self-describe	0	Postgraduate	70
35–44	65	Prefer not to say	2	Prefer not to say	2
45–54	7	Household income (€)		Food preferences	
55–64	4	Under 10,000	11	Vegan	2
65+	6	10,000-20,000	18	Vegetarian	5
		20,000–30,000	16	Omnivorous	97
		30,000–40,000	10	Gluten-free	2
		40,000–50,000	10	Organic	1
		50,000+	31	Pescatarian	1
		Prefer not to say	19	Carnivorous	7

Table 6.2 Comparison of dietary groups across various categories

Category	Vegetarians (incl. vegans)	Omnivorous	Carnivorous
Age groups			
18–24	28.57%	6.19%	0%
25–34	14.29%	23.71%	14.29%
35–44	28.57%	56.70%	57.14%
45–54	0%	7.22%	0%
55–64	0%	3.09%	14.29%
65+	28.57%	3.09%	14.29%
Gender distribution			
Male	28.57%	37.11%	71.43%
Female	71.43%	60.82%	28.57%
Non-binary	0%	2.06%	0%
Prefer not to say	0%	2.08%	0%
BMI (Body Mass Index)			
Underweight	0%	2.06%	0%
Normal weight	57.14%	63.92%	28.57%
Overweight	42.86%	27.84%	71.43%
Obese	0%	6.18%	0%
Environmental impact			
Very low	0%	13.40%	14.29%
Low	0%	17.53%	28.57%
Moderate	42.86%	47.42%	42.86%
High	42.86%	19.59%	14.29%
Very high	14.29%	2.06%	0%
Health concerns			
Very low	0%	4.12%	0%
Low	0%	7.22%	42.86%
Moderate	57.14%	38.14%	42.86%
High	28.57%	42.27%	14.29%
Very high	14.29%	8.25%	0%
Ethical concerns			
Yes	71.43%	27.08%	0%
No	28.57%	72.92%	100%

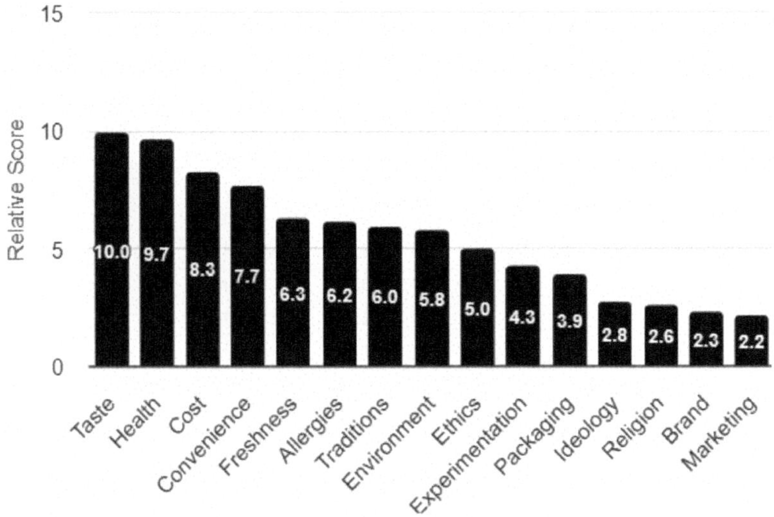

Fig. 6.3 Key factors affecting food habits of dietary groups

vegan groups, suggesting likely gender-related tendencies and cultural influences on dietary choices.

A wide variety of factors that influence people's choices of food were taken into consideration. Taste was considered the most important factor, closely followed by health, cost, and convenience. In addition to these, other factors like ingredient freshness, allergies, traditions and environmental concerns were also identified as significant in influencing food habits as shown in Fig. 6.3.

In addition, the analysis shows a higher proportion of overweight in the carnivorous group than their vegetarian and vegan counterparts. In this case, statistical analysis is supported by various research works, which often refer to plant-based diets as factors that promote lower body mass indices (BMIs) and better health outcomes [41].

6.3.2 VR Demonstration

Following the initial questionnaire, three participants, each representing a primary dietary group identified in the study, were invited to participate in the validation experiment within the VR environment where their visual attention points were observed. Table 6.3 displays the eye-tracking data results for these three individuals across six distinct food scenarios.

The results indicate that vegetarians have a unique preference for sustainable food situations showing increased preference for vegetables-based food scenarios. This finding is in harmony with their dietary standards and consciousness of the environ-

Table 6.3 Results of count, total, max, and average gaze duration per food scenario at a sampling ratio of $f = 50$ Hz and minimum gaze fixation$>$ 0.15 s

Group	Scenario	Count	Total gaze (s)	Avg gaze (s)	Max gaze (s)
Vegetarian	Fish	13	9.59	0.74	2.44
	Pasta	14	10.56	0.75	1,68
	Pizza	15	10.06	0.67	2.44
	Steak	8	3.48	0.44	0.74
	Fruits	11	10.34	0.94	0.74
	Salad	18	16.17	0.9	2.34
Omnivorous	Fish	10	4.00	0.40	0.88
	Pasta	2	1.08	0.54	0.58
	Pizza	8	4.12	0.52	1.04
	Steak	11	11.96	1.09	3.32
	Fruits	7	4.58	0.65	1.88
	Salad	3	2.02	0.67	0.90
Carnivorous	Fish	5	1.94	0.39	0.78
	Pasta	13	13.28	1.02	4.48
	Pizza	12	24.84	2.07	9.50
	Steak	5	12.48	2.50	6.28
	Fruits	6	3.96	0.66	1.34
	Salad	3	2.22	0.74	1.56

ment. As expected, those who self-identify as carnivorous showed a clear inclination towards the meat-based scenario, a choice known to be not environmentally friendly. On the other hand, participants who did not have any specific dietary preferences demonstrated interest in various other scenarios– even those considered less sustainable. The above findings yield the obvious result, confirming the expected alignment between dietary preferences and dietary groups.

Figure 6.4 shows heatmap results that have been normalized using a logarithmic transformation for a participant with a carnivorous diet, emphasizing more frequent gaze points on steak and pizza compared to other, more sustainable food options presented. The logarithmic transformation helps to minimize data skewness, improving the clarity and contrast in the heatmap, particularly useful when there's a broad variation in gaze counts.

To obtain insights about the visual attention points, heatmaps were created in VR space for a carnivorous viewer, depicting the particular zones that attracted the most visual attention. An example of these heatmaps is shown in Fig. 6.5.

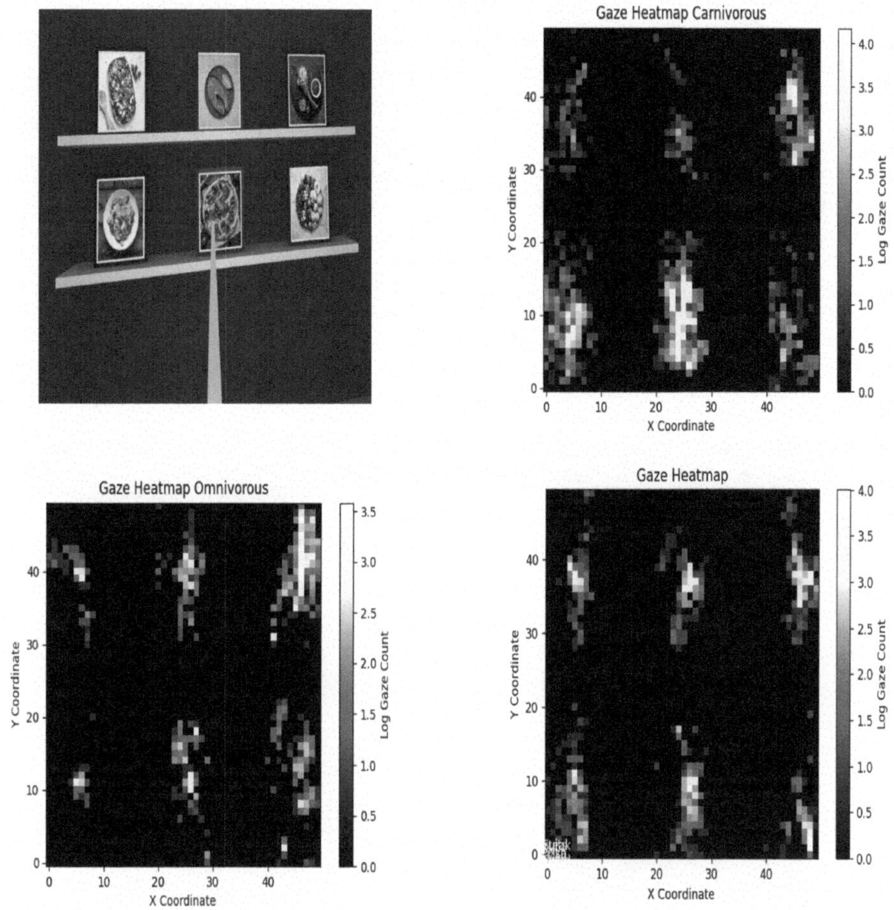

Fig. 6.4 Top left: Game view interaction points. Top right: Heatmaps for carnivorous participant. Bottom left: Heatmaps for omnivorous participant. Bottom right: Heatmaps for vegetarian participant

6.3.3 MR Demonstration

The development of immersive content takes place in an MR environment which aims at studying correlations between participants' eye-gazes. As part of this analysis, tailored data-storytelling elements are introduced to ensure a more engaging experience for the users as shown in Figs. 6.6 and 6.7. Nutritional values of the main ingredients from the food scenarios were calculated using NRF9.3 methodology [42] while CO2 emissions, water and land use estimates were taken from previous studies [43]. In addition, the setup tracks all the user interactions (eye tracking, user clicks etc.) with the key factors affecting food habits.

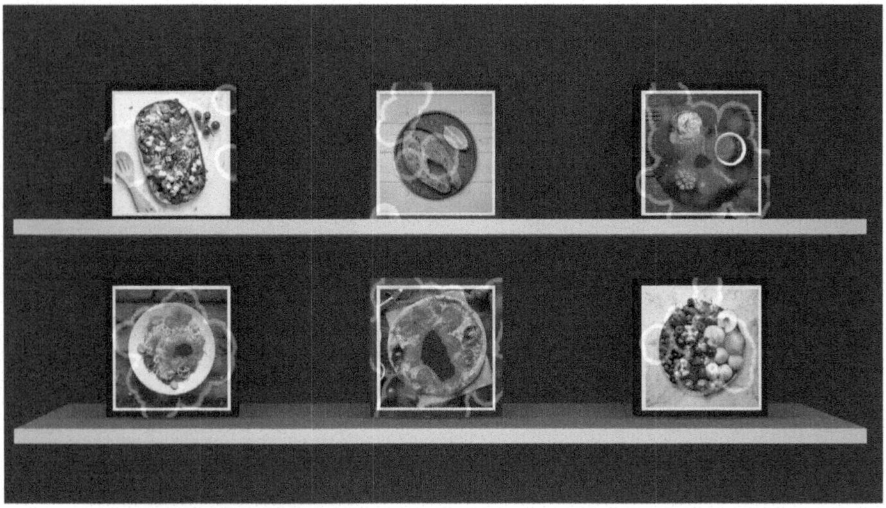

Fig. 6.5 Visual attention heatmaps generated within the VR environment for a carnivorous viewer

Fig. 6.6 Mixed reality environment setting

Table 6.4 presents data on the number of user clicks, distinct gazes, and the total, average, and maximum durations of gazes for various key metrics across all food scenarios for carnivorous, vegetarian, and omnivorous participants.

The data reveals that the omnivorous participant exhibited greater engagement with the presented factors, as indicated by a higher number of gazes and longer total

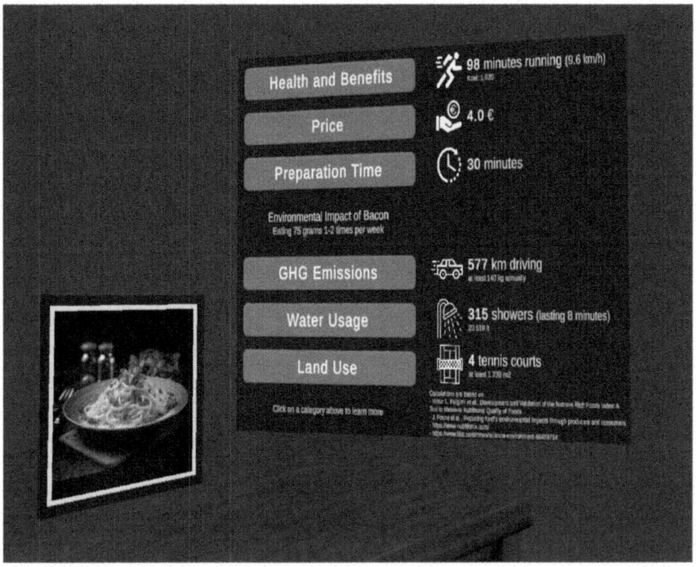

Fig. 6.7 Data storytelling visuals in MR

Table 6.4 Number of user clicks, number of Gazes, total, average and max Gaze duration on several key metrics for all food scenarios

Group	Factor	Clicks	Count	Total (s)	Avg (s)	Max (s)
Carnivorous	Health	20	30	9.70	0.32	0.68
	Water Use	19	69	26.24	0.38	1.58
	Price	19	65	28.58	0.44	1.16
	Land	12	60	20.78	0.34	1.28
	GHG	20	25	9.46	0.38	1.34
	Convenience	12	47	20.44	0.44	4.58
Vegetarian	Health	9	7	1.64	0.23	0.36
	Water Use	4	14	3.60	0.26	0.52
	Price	4	7	2.24	0.32	0.80
	Land	3	18	4.46	0.25	0.38
	GHG	5	17	3.88	0.23	0.36
	Convenience	5	25	8.28	0.33	1.00
Omnivorous	Health	11	69	41.08	0.59	2.88
	Water Use	10	66	35.60	0.54	3.64
	Price	13	69	38.72	0.56	3.46
	Land	10	59	33.86	0.57	2.80
	GHG	10	64	34.16	0.53	3.58
	Convenience	12	29	10.08	0.35	0.68

Table 6.5 Results of number of Gazes, total, max, and average Gaze duration per food scenario at a sampling ratio of $f = 50$ Hz and minimum gaze fixation > 0.15 s

Group	Scenario	Count	Total Gaze (s)	Avg (s)	Max (s)
Vegetarian	Carbonara	2	0.42	0.21	0.26
	Lentil soup	24	9.26	0.39	1.36
	Butter	10	3.72	0.37	0.88
	Beef stake	13	8.06	0.62	1.44
	Bean soup	5	1.62	0.32	0.52
	Buratta salad	31	17.06	0.55	3.48
Omnivorous	Carbonara	39	26.54	0.68	4.22
	Lentil soup	17	21.10	1.24	5.28
	Butter	14	8.76	0.63	2.14
	Beef stake	21	12.68	0.60	1.78
	Bean soup	17	8.76	0.52	2.24
	Buratta salad	22	17.94	0.82	3.46
Carnivorous	Carbonara	29	14.16	0.48	1.80
	Lentil soup	21	11.24	0.54	2.42
	Butter	24	7.60	0.32	1.02
	Beef stake	41	19.28	0.47	1.76
	Bean soup	25	12.24	0.49	1.20
	Buratta salad	38	17.66	0.46	1.76

and maximum gaze durations. This could be attributed to the complexity and variety of their diet choices, suggesting a potentially significant concern for the sustainability of their dietary habits. Additionally, the carnivorous participant's gazes in the MR environment were primarily focused on content featuring meat scenarios, more so than the vegetarian participants, as detailed in Table 6.5. This indicates that the MR environment successfully engages users in line with their dietary preferences as reflected in the scenarios displayed.

To enhance understanding of visual attention patterns, heatmaps were created within the MR environment for an omnivorous participant. These heatmaps highlight the specific areas that accumulated the most visual attention, as shown in Fig. 6.8.

Results show that the carnivorous participant demonstrated increased engagement with food consumption scenarios that are less healthy and sustainable, aligning with their dietary preferences. In this case, there was a significant interest in the health, greenhouse gas (GHG) emissions, and water use metrics associated with their food choices. This finding suggests a potential opportunity to educate such dietary groups about the environmental impacts of their choices in MR environments.

To further understand how various dietary groups prioritize key factors across specific food scenarios, this study examines interactions with a particular emphasis on meat consumption as shown in Table 6.6.

It is revealed that the carnivorous participant demonstrated the most significant interaction with the meat-based scenario in terms of gaze duration and frequency,

Fig. 6.8 Visual attention heatmaps generated within the MR environment for omnivorous participant

Table 6.6 Analysis of user clicks, Gaze count, total, average, and maximum Gaze duration for key metrics for a meat-based scenario

Group	Factor	Clicks	Count	Total (s)	Avg (s)	Max (s)
Carnivorous	Health	4	3	1.00	0.33	0.52
	Water use	6	27	11.44	0.42	1.58
	Price	4	15	7.04	0.47	1.16
	Land use	4	19	6.74	0.36	1.28
	GHG emissions	7	12	3.84	0.32	0.82
	Convenience	2	15	5.22	0.35	0.74
Vegetarian	Health	1	1	0.16	0.16	0.16
	Water Use	1	1	0.52	0.52	0.52
	Price	0	0	0	0	0
	Land use	1	2	0.42	0.22	0.22
	GHG emissions	1	1	0.36	0.36	0.36
	Convenience	1	3	0.64	0.21	0.28
Omnivorous	Health	1	3	2.58	0.86	1.60
	Water Use	1	4	5.20	1.30	3.64
	Price	2	12	9.34	0.78	3.46
	Land use	1	3	3.36	1.12	2.46
	GHG emissions	1	3	4.78	1.59	3.58
	Convenience	1	5	1.80	0.36	0.54

indicating a strong focus as health, water use, and GHG emissions factors. On the other hand, as expected the vegetarian participants showed significantly lower engagement in the meat-based scenarios, with minimal clicks and gaze counts. In the case of the omnivorous participant, a balanced interaction across all metrics is found, with their gaze durations and counts being consistently moderate, revealing an evenly distributed interest in the factors influencing meat consumption.

A significant challenge addressed in this study was configuring the MR scenarios to allow participants to interact with a wide range of food options reflecting diverse dietary preferences. To ensure the precision of this immersive experience, eye-tracking technology integrated into the MR setup was utilized to track participants' visual engagement with key elements influencing their food choices. Analyzing the eye-tracking data provided clear insights into individual preferences and visual biases toward specific food categories and ingredients, which further informed the understanding of self-reported eating behaviors in relation to actual perceptions within the MR environment.

6.4 Conclusions and Future Studies

This study validates a flexible and adaptable methodology of using both VR and MR to assess dietary habits through the incorporation of eye-tracking technology and immersive data storytelling visuals. Preliminary results demonstrated interesting insights into consumer behaviour. In the VR environment, carnivorous participants displayed a greater interest in meat-based scenarios, while vegetarians were more engaged with sustainable food options. Similarly, the MR assessment indicated that individuals with sustainable dietary habits tended to spend more time viewing images of sustainable food options, showing an increased interest compared to those with less sustainable diets.

These findings provide preliminary evidence that VR and MR environments can be effective tools for studying dietary choices. Analyses revealed that those with carnivorous diets focused predominantly on meat scenarios, whereas vegans showed greater interest in more sustainable options. However, to arrive at more comprehensive conclusions, further research is required, incorporating qualitative metrics such as selection patterns and physiological responses (pupil dilation, heart rate) and underscoring the potential of mixed reality assessments to not only assessing user interest but also influence behavioral shifts toward healthier and more sustainable eating habits.

Additional MR scenarios will be developed to encompass a broader spectrum of consumer behaviors and habits, such as circular economy and climate resilience, with further data collection aimed at evaluating how targeted campaigns can promote a transition to more sustainable practices. These enhanced insights will also explore how small-scale actions can promote societal acceptance and a positive shift in sustainable habits, thereby identifying which actions and ultimately policies are most effective.

Acknowledgements This research is supported by Funding from the EU Horizon 2020 project under grant agreement no. 101081617 (CHOICE).

References

1. K. Kim, J. Choi, J. Chin, B. Lee, D. Na, Eye-Tracking Metrics for Figure-Copying Processes in Early- vs. Late-Onset Alzheimer's Disease. Frontiers in Neurology, 13. https://doi.org/10.3389/fneur.2022.844341 (2022)
2. C. Wu, J.S. Cha, J.E. Sulek, T. Zhou, C. Sundaram, J. Wachs, D. Yu, Eye-Tracking Metrics Predict Perceived Workload in Robotic Surgical Skills Training. Human Factors, 62, 1365-1386. https://doi.org/10.1177/0018720819874544 (2019)
3. T. Morimoto, T. Kobayashi, H. Hirata, K. Otani, M. Sugimoto, M. Tsukamoto, T. Yoshihara, M. Ueno, M. Mawatari, XR (Extended Reality: Virtual Reality, Augmented Reality, Mixed Reality) Technology in Spine Medicine: Status Quo and Quo Vadis. J. Clin. Med. **11**(2022). https://doi.org/10.3390/jcm11020470
4. E. Markopoulos, M. Luimula, C. Calbureanu-Popescu, P. Markopoulos, P. Ranttila, S. Laukkanen, N. Laivuori, W. Ravyse, J. Saarinen, T. Nghia, Neural Network Driven Eye Tracking Metrics and Data Visualization in Metaverse and Virtual Reality Maritime Safety Training. https://api.semanticscholar.org/CorpusID:251465705 (2022)
5. S. Mystakidis, E. Berki, J. Valtanen, Deep and Meaningful E-Learning with Social Virtual Reality Environments in Higher Education: A Systematic Literature Review. Applied Sciences. https://doi.org/10.3390/APP11052412 (2021)
6. N. Kim, H. Lee, Assessing Consumer Attention and Arousal Using Eye-Tracking Technology in Virtual Retail Environment. Front. Psychol. **12**, 665658 (2021). https://doi.org/10.3389/fpsyg.2021.665658
7. A.T. Duchowski, Eye Tracking Methodology. Springer-Verlag, Londyn. (2007). https://doi.org/10.1007/978-3-319-57883-5
8. J. Clement, Visual influence on in-store buying decisions: an eye-track experiment on the visual influence of packaging design. Journal of Marketing Management, 23(9–10), pp. 917–928. https://doi.org/10.1362/026725707x250395 (2007)
9. J. Clement, T. Kristensen, K. Grønhaug, Understanding consumers' in-store visual perception: The influence of package design features on visual attention. Journal of Retailing and Consumer Services, 20(2), pp. 234–239. https://doi.org/10.1016/j.jretconser.2013.01.003 (2013)
10. S. Atalay, H. Bodur, D. Rasolofoarison, Shining in the Center: Central Gaze Cascade Effect on Product Choice. Journal of Consumer Research. https://doi.org/10.2139/ssrn.2258531 (2012)
11. P. Huddleston, B.K. Behe, S. Minahan, R.T. Fernandez, Seeking attention: An eye tracking study of in-store merchandise displays. International Journal of Retail and Distribution Management, 43(6), pp. 561–574. https://doi.org/10.1108/IJRDM-06-2013-0120 (2015)
12. P. Chandon, J.W. Hutchinson, E.T. Bradlow, S.H. Young, Does In-Store Marketing Work? Journal of Marketing, 73(6). https://omura.wordpress.com/2009/10/31/does-in-store-marketing-work/ (2009)
13. N.M. Gunaratne, S. Fuentes, T.M. Gunaratne, D.D. Torrico, H. Ashman, C. Francis, C. Gonzalez Viejo, F.R. Dunshea, Consumer Acceptability, Eye Fixation, and Physiological Responses: A Study of Novel and Familiar Chocolate Packaging Designs Using Eye-Tracking Devices. Foods 2019, 8, 253. https://doi.org/10.3390/foods8070253
14. E. Wästlund, T. Otterbring, A. Gustafsson, P. Shams, Heuristics and resource depletion: Eye-tracking customers' in situ gaze behavior in the field. Journal of Business Research, 68(1), pp. 95–101. https://doi.org/10.1016/j.jbusres.2014.05.001 (2015)
15. T.S. Teixeira, M. Wedel, R. Pieters, Moment-to-Moment Optimal Branding in TV Commercials: Preventing Avoidance by Pulsing. Mark. Sci. **29**(5), 783–804 (2010). https://doi.org/10.1287/mksc.1100.0567

16. A.R. Smink, E.A. van Reijmersdal, S.C. Boerman, Effects of Brand Placement Disclosures: An Eye Tracking Study Into the Effects Of Disclosures and the Moderating Role of Brand Familiarity. Advances in Advertising Research, 8, pp. 85–96. Springer Nature, Wiesbaden. https://doi.org/10.1007/978-3-658-18731-6_5 (2017)
17. I. Puškarević, U. Nedeljković, V. Dimovski, K. Možina, Eye tracking study of attention to print advertisements: Effects of typeface figuration. Journal of Eye Movement Research, 9(5). https://doi.org/10.16910/jemr.9.5.6 (2016)
18. S.B. Hutton, S. Nolte, The effect of gaze cues on attention to print advertisements. Applied Cognitive Psychology, 25(6), pp. 887–892. https://doi.org/10.1002/acp.1763 (2011)
19. M. Porta, A. Ravarelli, F. Spaghi, Online newspapers and ad banners: An eye tracking study on the effects of congruity. Online Inf. Rev. **37**(3), 405–423 (2013). https://doi.org/10.1108/OIR-01-2012-0001
20. A.M. Barreto, Do users look at banner ads on Facebook? Journal of Research in Interactive Marketing, 7(2), pp. 119–139. https://doi.org/10.1108/jrim-mar-2012-0013 (2013)
21. M. Resnick, W. Albert, The Impact of Advertising Location and User Task on the Emergence of Banner Ad Blindness: An Eye-Tracking Study. International Journal of Human-Computer Interaction, 30(3), pp. 206–219. https://doi.org/10.1080/10447318.2013.847762 (2014)
22. J. Lee, J.-H. Ahn, 'Attention to Banner Ads and Their Effectiveness: An Eye-Tracking Approach', International Journal of Electronic Commerce, 17(1), pp. 119–137. https://doi.org/10.2753/JEC1086-4415170105 (2012)
23. D. Cosic, Neuromarketing in Market Research. Interdisciplinary Description of Complex Systems, 14(2), pp. 139–147. https://doi.org/10.7906/indecs.14.2.3 (2016)
24. A. Pai, A. Sabharwal, Food Habits: Insights from Food Diaries via Computational Recurrence Measures. Sensors (Basel, Switzerland), 2022, Vol. 22, https://doi.org/10.3390/s22072753 (2022)
25. M. Svendsen, S. Tonstad, Accuracy of food intake reporting in obese subjects with metabolic risk factors. Br. J. Nutr. **95**(3), 640–9 (2006). https://doi.org/10.1079/bjn20051662. (PMID: 16512951)
26. V. Sember, K. Meh, M. Sorić, G. Starc, P. Rocha, G. Jurak, Validity and Reliability of International Physical Activity Questionnaires for Adults across EU Countries: Systematic Review and Meta Analysis. International Journal of Environmental Research and Public Health, 17. https://doi.org/10.3390/ijerph17197161 (2020)
27. S. Faghih, P. Keshani, A. Salar, S.H. Rajaei, Z. Mirzaei, S.M. Moosavi, Z. Hematdar, Assessment of Obesity, Unhealthy Food Habits, and Nutritional Knowledge of Primary School Children. International Journal of School Health, https://doi.org/10.17795/INTJSH-25186 (2015)
28. G.M. Manzoni, G.L. Cesa, M. Bacchetta, G. Castelnuovo, S. Conti, A. Gaggioli, G. Riva, Virtual Reality-Enhanced Cognitive-Behavioral Therapy for Morbid Obesity: A Randomized Controlled Study with 1 Year Follow-Up. Cyberpsychology, Behavior, and Social Networking, 19(2), 134-140. https://doi.org/10.1089/cyber.2015.0208 (2016)
29. A. Plechatà, T. Morton, F. Perez-Cueto, G. Makransky, A randomized trial testing the effectiveness of virtual reality as a tool for pro-environmental dietary change. Scientific Reports, 12. https://doi.org/10.1089/cyber.2015.0208 (2022)
30. T. Ruppenthal, Eye-Tracking Studies on Sustainable Food Consumption: A Systematic Literature Review. Sustainability. https://doi.org/10.3390/su152316434 (2023)
31. G. Benvegnù, A. Piva, C. Cadorin, V. Mannari, M. Girondini, A. Federico, S. Tamburin, C. Chiamulera, The effects of virtual reality environmental enrichments on craving to food in healthy volunteers. Psychopharmacology, 241, 49-60. https://doi.org/10.1007/s00213-023-06462-z (2023)
32. M. Allman-Farinelli, K. Ijaz, H. Tran, H. Pallotta, S. Ramos, J. Liu, L. Wellard-Cole, R. Calvo, A Virtual Reality Food Court to Study Meal Choices in Youth: Design and Assessment of Usability. JMIR Formative Research, 3. https://doi.org/10.2196/12456 (2019)
33. F. Ramousse, P. Raimbaud, P. Baert, C. Helfenstein-Didier, A. Gay, C. Massoubre, B. Galusca, G. Lavoué, Does this virtual food make me hungry? Effects of visual quality and food type in virtual reality. Frontiers Virtual Real., 4. https://doi.org/10.3389/frvir.2023.1221651 (2023)

34. E. Smit, M. Meijers, L.N. van der Laan, Using Virtual Reality to Stimulate Healthy and Environmentally Friendly Food Consumption among Children: An Interview Study. International Journal of Environmental Research and Public Health, https://doi.org/10.3390/ijerph18031088 (2021)
35. S.M. Max, K. Schag, K. Giel, C. Plewnia, Behavioural biases in the interaction with food objects in virtual reality and its clinical implication for binge eating disorder. Eating and Weight Disorders, 28. https://doi.org/10.1007/s40519-023-01571-2 (2023)
36. P. Schroeder, E. Collantoni, J. Lohmann, M.V. Butz, C. Plewnia, Faster Grasping of High-Calorie Food Objects in Virtual Reality. https://doi.org/10.21203/rs.3.rs-524244%2Fv1 (2021)
37. P. Rauschnabel, R. Felix, C. Hinsch, H. Shahab, F. Alt, What is XR? Towards a Framework for Augmented and Virtual Reality. Computers in Human Behavior, 133.https://doi.org/10.1016/j.chb.2022.107289 (2022)
38. V. Sundstedt, V. Garro, A Systematic Review of Visualization Techniques and Analysis Tools for Eye-Tracking in 3D Environments. Front. Neuroergon. **3**, 910019 (2022). https://doi.org/10.3389/fnrgo.2022.910019
39. A.T. Duchowski, *Eye Tracking Methodology: Theory and Practice* (Eye Tracking Methodology, Springer (Theory and Practice, 2017)
40. K. Holmqvist, N. Nyström, R. Andersson, R. Dewhurst, H. Jarodzka, J. Van de Weijer (eds.), *Eye tracking: a comprehensive guide to methods and measures* (Oxford University Press, Oxford, UK, 2011)
41. J.J. Ferguson, C. Oldmeadow, G.D. Mishra, M.L. Garg, Plant-based dietary patterns are associated with lower body weight, BMI and waist circumference in older Australian women. Public Health Nutr. **25**(1), 18–31 (2022). https://doi.org/10.1017/S1368980021003852
42. D. Sluik, M.T. Streppel, L. van Lee, A. Geelen, E.J. Feskens, Evaluation of a nutrient-rich food index score in the Netherlands. J Nutr Sci. **20**(4), e14 (2015). https://doi.org/10.1017/jns.2015.4. PMID: 26097700; PMCID: PMC4462757
43. J. Poore, T. Nemecek, Reducing food's environmental impacts through producers and consumers, Science, Vol 360, Issue:6392, pp. 987-992, https://doi.org/10.1126/science.aaq0216

Chapter 7
Towards Hybrid Intelligence in Journalism: Findings and Lessons Learnt from a Collaborative Analysis of Greek Political Rhetoric by ChatGPT and Humans

Thanasis Troboukis, Kelly Kiki, Antonis Galanopoulos, Pavlos Sermpezis, Stelios Karamanidis, Ilias Dimitriadis, and Athena Vakali

Abstract This chapter introduces a research project titled "Analyzing the Political Discourse: A Collaboration Between Humans and Artificial Intelligence", which was initiated in preparation for Greece's 2023 general elections. The project focused on the analysis of political leaders' campaign speeches, employing Artificial Intelligence (AI), in conjunction with an interdisciplinary team comprising journalists, a political scientist, and data scientists. The chapter delves into various aspects of political discourse analysis, including sentiment analysis, polarization, populism, topic detection, and Named Entities Recognition (NER). This experimental study investigates the capabilities of large language model (LLMs), and in particular OpenAI's ChatGPT, for analyzing political speech, evaluates its strengths and weaknesses, and highlights the essential role of human oversight in using AI in journalism projects

T. Troboukis · K. Kiki (✉)
Incubator for Media Education and Development (iMEdD), Athens, Greece
e-mail: k.kiki@imedd.org

T. Troboukis
e-mail: troboukis@gmail.com

A. Galanopoulos
School of Political Sciences, Aristotle University of Thessaloniki, Thessaloniki, Greece
e-mail: antonisgal@gmail.com

P. Sermpezis · S. Karamanidis · I. Dimitriadis · A. Vakali
School of Informatics, Aristotle University of Thessaloniki, Thessaloniki, Greece
e-mail: sermpezis@csd.auth.gr

S. Karamanidis
e-mail: skaraman@csd.auth.gr

I. Dimitriadis
e-mail: idimitriad@csd.auth.gr

K. Karpouzis and Y. Skarpelos (eds.), *Computational Methods for the Digital Humanities*, Learning and Analytics in Intelligent Systems 57,
https://doi.org/10.1007/978-3-032-05099-1_7

and potentially other societal sectors.The project stands as an innovative example of human-AI collaboration (known also as "hybrid intelligence") within the realm of digital humanities, offering valuable insights for future initiatives.

Keywords Political discourse analysis · Elections · Artificial intelligence · ChatGPT · Sentiment analysis · Topic detection · Polarization · Populism · Data journalism

7.1 Introduction

The incorporation of Artificial Intelligence (AI)[1] into the practice of journalism presents potential advantages [2] in terms of improving the efficiency and excellence of news creation and distribution, as well as bolstering trustworthiness [22]. The proliferation of AI tools and open datasets made available to journalists to incorporate new (data and AI driven) approaches in their projects [3, 4, 7, 12, 13], is a trend that is only expected to increase in the following years [6].

Advanced AI algorithms, including Natural Language Processing (NLP) models and Large Language Models (LLMs), such as ChatGPT [23], have demonstrated competence in automating labor-intensive activities [8, 11], including data mining, sentiment analysis [1, 31], basic reporting, annotating misinformation [1, 14, 26] and hate speech [15], etc.. The implementation of automation not only accelerates the pace of news dissemination but also presents opportunities for the field of Investigative Journalism [25, 29], allowing human journalists to allocate their efforts toward more intricate analytical and investigative endeavors. Moreover, the ability of AI to analyze extensive datasets can provide a unique understanding of public opinion, political trends [17, 21, 30], and emerging societal concerns. Therefore, this can provide a richer, more nuanced backdrop for journalists to construct their narratives.

Nevertheless, it is imperative to acknowledge that AI ought to function as a instrument to human expertise, rather than a substitute, due to its constraints in comprehending contextual nuances and its propensity for innate biases [5, 20]. Therefore, although AI has the potential to significantly transform the field of journalism, its successful incorporation necessitates careful monitoring and ethical deliberation.

---

[1] While AI and its concepts are widely known, to ensure a common understanding among all readers of this article, we state the following definitions [2]:

- *Artificial intelligence (AI)* is a collection of ideas, technologies, and techniques that relate to a computer system's capacity to perform tasks normally requiring human intelligence.
- *Generative AI* is a subfield of AI in its own right,that involves the generation of new data, such as text, images, or code, based on a given set of input data. ChatGPT [23] is a prominent example of generative AI.

Motivation and Goals: Motivated by (i) the emerging capabilities afforded by AI and (ii) the journalistic need to timely report on a political topic, we launched the project "Analyzing the political discourse: a collaboration between humans and Artificial Intelligence" [16] in spring 2023, ahead of the general elections in Greece in May of the same year.

The project aimed to achieve two primary objectives:

First, to perform a comprehensive examination of political leaders' campaign speeches, with the aim of uncovering qualitative features of their discourse that go beyond the usual sound bite excerpts presented in the news cycle, such as: identify the main themes that political leaders emphasize in their campaign speeches, as well as to analyze the sentiments conveyed. Additionally, the study quantified the extent to which elements of polarization and populism are present in the leaders' rhetoric.

Second, to explore the boundaries of modern technology and underscore the LLMs proficiency in conducting qualitative and comparative research analysis on political discourse. What are the potential, challenges, and limitations of using AI in journalism projects? How can we craft robust and reliable methodologies? Is human supervision still needed and to what extent?

Contributions: Overall, the project was designed as an experimental collaboration between an interdisciplinary team of individuals, including (data) journalists, a political scientist, computer and data scientists, and web developers, along with AI. This "hybrid intelligence" led to the following contributions.

Methodology and tools: We collected a large dataset of 171 speeches of the leaders of political parties (Sect. 7.2.1). We crafted a methodology based on journalism and political science methods to analyze the characteristics of the political rhetoric, and automated the analysis using ChatGPT (Sect. 7.2.2), followed by a human validation of the results (Sect. 7.2.3); in total, leading to a "hybrid intelligence" approach. Finally, we created an informative online tool with insightful interactive visualizations (Sect. 7.2.4) useful to journalists, researchers, and the general public.

While our use case (dataset) was on texts in Greek, our methodology is applicable to any language. To further encourage its use by other journalists, we open-source the data, prompting code and visualizations (see details in [18]).

Analysis of political rhetoric during the Greek national elections: During our project, and while the pre-election period was still ongoing, we published the results of the analysis[2] through our online tool (Sect. 7.2.4), offering thus a continuous monitoring of the political rhetoric during the elections. This enabled also a number of timely political analyses and articles by journalists, political scientists, academics, etc. (available at [16]). We discuss the main findings of the political analyses in Sect. 7.3.

Evaluation of ChatGPT's efficiency for journalism projects: Our hybrid intelligence approach (Sects. 7.2.2 and 7.2.3) enabled us to evaluate to what extent ChatGPT was efficient in answering our journalistic questions. To this end, we provide a detailed analysis of the accuracy results in Sect. 7.4.

[2] For each speech, the analysis results were made available within one day of its occurrence.

Our findings show that ChatGPT demonstrated exceptional proficiency in sentiment analysis and Named Entity Recognition (NER) when applied to the talks of political leaders. Furthermore, it demonstrated efficacy in the detection of the topics that were given in the realm of public discourse. Nevertheless, prior to publication, the data necessitated significant human vetting. ChatGPT exhibited notable bias in its ability to identify discourse associated with polarization and populism.

Summarizing, the incorporation of ChatGPT into our operational framework significantly expedited the examination of political leaders' speeches and the dissemination of our research outcomes, occasionally diminishing the full procedure to a little three-hour timeframe. Nevertheless, the presence of inherent biases within the model, particularly towards left-wing rhetoric, necessitated the implementation of stringent human oversight.

Up to now, conversations about AI in journalism have mostly centered on automating reporting tasks or news delivery. However, this study suggests that journalism should broaden its scope to explore AI's potential in other aspects of the field. For example, AI could revolutionize political reporting by offering data-driven insights that might escape human analysis. This would allow journalists to provide their audience with more fact-based and in-depth examinations of complex, often theoretical, issues.

7.2 Methodology

In this section, we describe the methodology[3] for collecting the texts of the speeches (Sect. 7.2.1), automatically analyzing them using ChatGPT (Sect. 7.2.2) and then manually validating the results (Sect. 7.2.3), and generating the visualizations (Sect. 7.2.4) that have been used in our project. Figure 7.1 depicts the overview of the methodology.

7.2.1 Collection and Processing of Speeches

The project's initiation was deliberately synchronized with the commencement of the first election period leading up to the parliamentary elections in Greece on May 21, 2023. Our primary focus was the analysis of campaign speeches delivered by the political leaders of the six parties that held seats in the Greek parliament during the most recent parliamentary term (18th Parliamentary Term, spanning from July 17, 2019, to April 22, 2023). These leaders include:

[3] We refer the interested reader to [18] for further details of the working methodology.

- *Kyriakos Mitsotakis*, Prime Minister and President of New Democracy
- *Alexis Tsipras*, President of SYRIZA
- *Nikos Androulakis*, President of PASOK-Movement for Change
- *Dimitris Koutsoumbas*, Secretary General of the Communist Party of Greece
- *Kyriakos Velopoulos*, President of Greek Solution
- *Yanis Varoufakis*, Secretary General of MeRA25

Following the May 2023 elections, a so-called "one-day parliament" was formed due to the inability to establish a government in light of the election results. Subsequently, parliamentary elections were called again. In the second phase of our project, during the analysis of campaign speeches leading up to the parliamentary elections on June 25, 2023, we focused on the leaders who were part of the initial study during the first campaign period and represented parties that entered the one-day parliament as a result of the May 2023 elections. This included all of the above leaders except for Y. Varoufakis.

For this study, any public speech delivered by the aforementioned political leaders in a public setting, in the presence of an audience, within the official pre-election period on each occasion, was considered a campaign speech. However, short statements or ad hoc talks (i.e., those lasting less than 15 minutes) and political leaders' interactions with citizens were excluded from the study. Only complete campaign speeches that were available in their entirety were subject to analysis. Due to the fact that K. Velopoulos did not deliver campaign speeches, as defined above, his opening remarks in various press conferences were analyzed. This exception should be noted, as the analysis of Velopoulos' discourse could not be directly compared to that of other political leaders.

To collect the campaign speeches, our primary sources were the press offices of the political parties, followed by the political parties' and their leaders' websites and/or social media accounts. We supplemented our sources list with contributions from the Athens-Macedonian News Agency (ANA-MPA), the Hellenic Broadcasting Corporation (ERT), and other local media outlets in cases where a speech was not fully available through the previously mentioned political parties' communication channels.

The analysis relied on the written text of each political leader's speech whenever available. If the text provided by the parties was already divided into paragraphs based on topics, it was retained in that format. If the written text was not already structured into paragraphs or was organized based on oral delivery, our team of journalists edited the text by segmenting it into paragraphs based on the main topic. In cases where the campaign speeches were not provided in written texts by the parties but were available in audiovisual formats, we used an AI-powered transcription tool[4] to convert the audio into text.

After collecting and/or formatting the written text of each campaign speech in Greek, we automatically translated it into English using a machine translation ser-

[4] Transkriptor, https://transkriptor.com/.

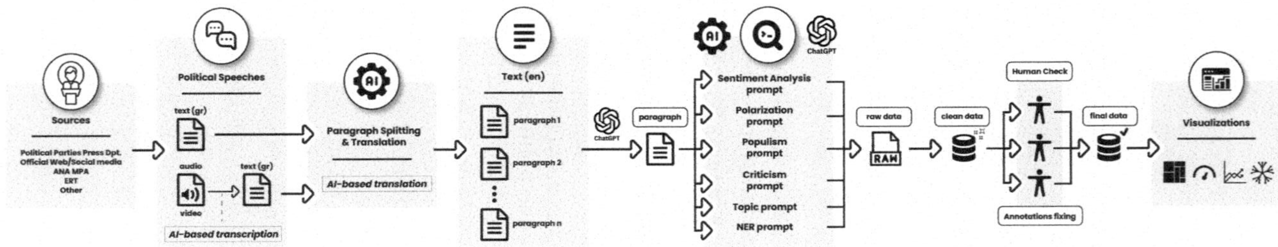

Fig. 7.1 Methodology of the hybrid intelligence political discourse analysis project

vice.[5] Subsequently, ChatGPT was tasked with analyzing this translated speech text, guided by definitions and context provided by the working group.

7.2.2 Speech Analysis and Prompting ChatGPT

The campaign speeches of political leaders were subjected to analysis at the paragraph level, with the aim of identifying whether the leader primarily focused on criticizing political opponents or presenting their party's agenda, the main topic/theme of the abstract, the dominant sentiment, levels of political polarization and populism detected, and any named entities.

Using an automated way, we prompted ChatGPT (the gpt-3.5-turbo model) to provide paragraph-based outputs for the following list of analyses:

- A value of either "criticism" or "political agenda" to indicate if the political leader primarily criticized opponents or referred to their party's ideas, opinions, positions, or program proposals.
- Identification of the most likely topic/theme discussed from a predefined list of 33 specific topics provided to ChatGPT by the working team (see [18] for the detailed list of topics).
- Assignment of a sentiment value on a scale of –1 to 1, indicating the negativity, neutrality, or positivity of the passage, to be later classified accordingly based on predefined scales by the working group.
- Assignment of a value on a scale of 0 to 1, indicating the level of political polarization within the passage, to be later classified as either no/low, medium, or high polarization based on predefined scales by the working group.
- Assignment of a value on the same scale to indicate the level of populism within the passage, to be later classified accordingly.
- Extraction of a list of named entities in the passage, categorized as individuals, groups, organizations, political parties, locations, countries, or dates.

Some of our prompts to ChatGPT were provided with carefully considered context, aiming to steer the model's responses by accounting for specific situations that would impact its output, with the goal of mitigating algorithmic bias. For example, when determining whether the text primarily criticized opponents or presented a political agenda and when identifying the main theme, we provided the name and status of the political leader. For instance, "you are reading a passage from a pre-election speech of the Greek Prime Minister, K. Mitsotakis, leader of the New Democracy political party" or "you are reading a passage from a pre-election speech of A. Tsipras, a Greek politician and leader of the opposition party SYRIZA (Coalition of the Radical Left)", was the context provided in cases involving Mitsotakis and Tsipras. To the contrary, we did not disclose the names of the political leaders

[5] DeepL, https://www.deepl.com/ through its API.

when evaluating the levels of polarization and populism, in order to bypass ChatGPT's inherent bias, especially towards left-wing rhetoric. Most importantly, when evaluating the levels of polarization and populism in a text, we didn't employ ChatGPT's pre-existing "knowledge"; to mitigate biases stemming from its training data, we provided definitions[6] of polarization and populism as drafted by the political scientist in the working group. This context remained consistent in all cases.

For each unique campaign speech, we compiled a dataset with rows corresponding to the number of paragraphs and columns representing the research variables. Following this, we conducted a human review of the ChatGPT results.

7.2.3 The Data Validation Process

For determining whether each speech passage focused on criticizing opponents or presenting a political agenda and identifying the major theme, data validation involved a minimum of two journalists who cross-verified the ChatGPT results. Although the analysis was conducted on text translated into English, validation was performed on the original Greek text. Any inaccuracies were rectified by the working group. Regarding sentiment analysis, polarization, and populism levels, two members of the research team–the political scientist and a journalist–reviewed ChatGPT's outputs. In cases of inaccuracies, the working group made the necessary corrections as follows:

Sentiment analysis: Each passage was initially categorized based on ChatGPT's assigned value: negative for scores between –1 and –0.34, neutral for scores between –0.33 and 0.33, or positive for scores between 0.34 and 1. If ChatGPT assigned a sentiment score that placed a paragraph into a different category than human judgment, that value was excluded and treated as missing data in subsequent analysis and relevant visualizations. This approach was adopted to prevent ad hoc human intervention in the sentiment indicator's average, which was used to classify the entire campaign speech as negative, neutral, or positive.

Levels of polarization and populism: Each passage was initially categorized based on ChatGPT's assigned value into one of three categories: zero/low level for scores between 0 and 0.5, medium level for scores between 0.51 and 0.8, or high level for scores between 0.81 and 1. If ChatGPT assigned a value that differed from human judgment, the political scientist, a member of the working team, adjusted this value

[6] We paid high attention to clearly define the two terms and avoid stereotypical approaches and normative bias. The definitions of polarization and populism provided as context to ChatGPT are found in the detailed published methodology [18]. For further information on relevant definitions, see [9, 10].

to 0,[7] 0.6, or 0.9, as appropriate for zero/low level, medium level or high level (of either polarization or populism) classification.

7.2.4 The Data Visualizations

The validated results were visualized with various charts that were constantly updated as the campaign period evolved. The major visualizations included:

- A treemap displaying the shares of topics covered by each political leader in their speeches. These percentages represented the estimated proportion of words spoken on each topic relative to the total number of words in the speech. An example is given in Fig. 7.2.
- A speedometer plotting the sentiment analysis of each speech (Fig. 7.3). The position of the needle indicated the average sentiment values at the paragraph level, leading to the classification of the speech as negative, neutral, or positive.

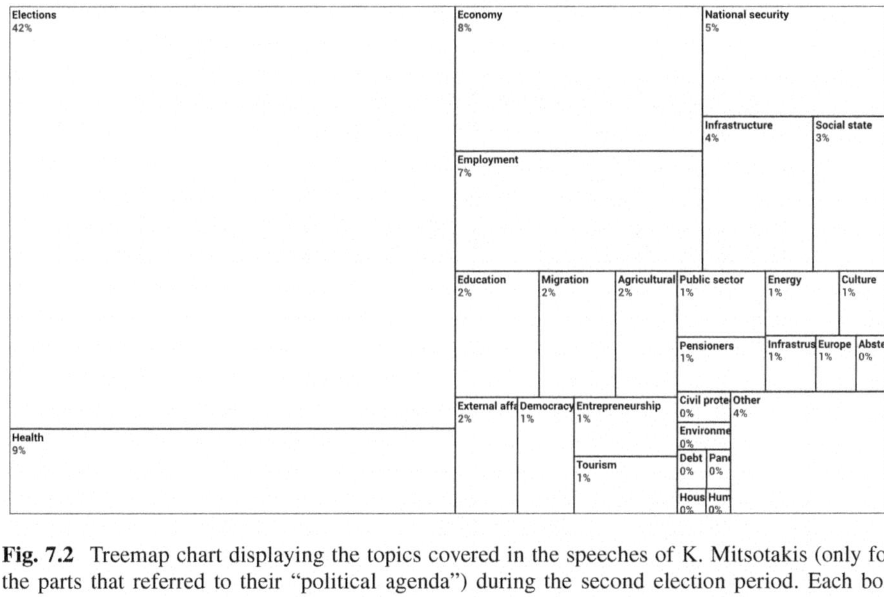

Fig. 7.2 Treemap chart displaying the topics covered in the speeches of K. Mitsotakis (only for the parts that referred to their "political agenda") during the second election period. Each box corresponds to a different topic, and its size is proportional to the percentage of paragraphs referring to this topic

[7] The majority of excerpts in the zero/low level category had scores very close to 0. To avoid the potential introduction of extreme values through human intervention within this category, a corrective value of 0 was established for this specific class. Mean scores were used as the corrective values for the remaining categories.

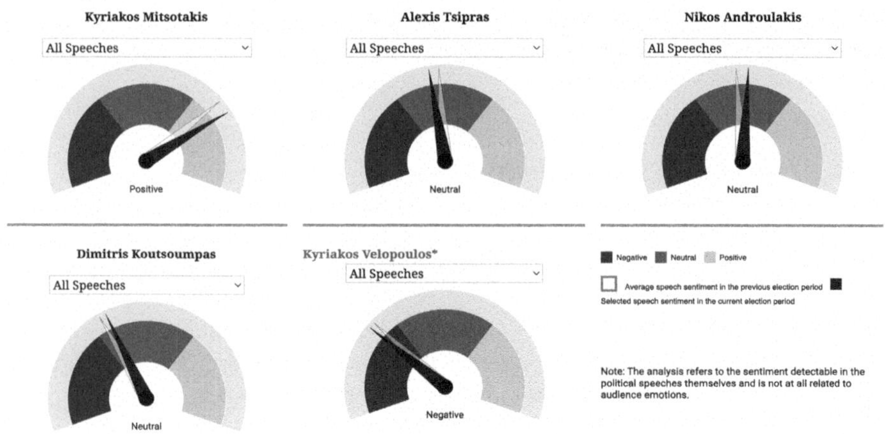

Fig. 7.3 Speedometer charts displaying the average sentiment for each political leader during the first (light gray) and second (dark gray) election periods. Dark gray/gray/light gray areas correspond to negative/neutral/positive sentiments, respectively

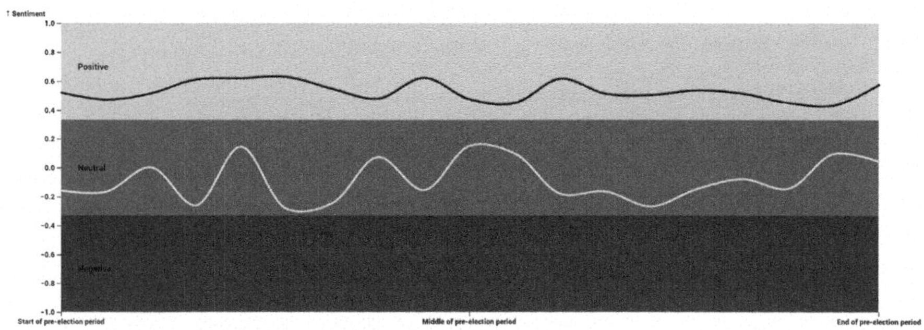

Fig. 7.4 Line chart depicting the sentiment evolution over the course of the second election period (x-axis) for K. Mitsotakis (black line) and A. Tsipras (white line). The y-axis corresponds to sentiment values, from –1 to 1, and colored areas to negative (dark gray), neutral (gray) and positive (light gray) sentiments

- Line charts depicting the sentiment evolution in specific speeches or over the course of the election period for two different leaders, based on the user's selection (Fig. 7.4).
- Step charts illustrating the evolution of polarization and populism levels during each speech or throughout a campaign period (Fig. 7.5).
- A heatmap representing the sentiment by theme discussed in each leader's speeches throughout the campaign period.

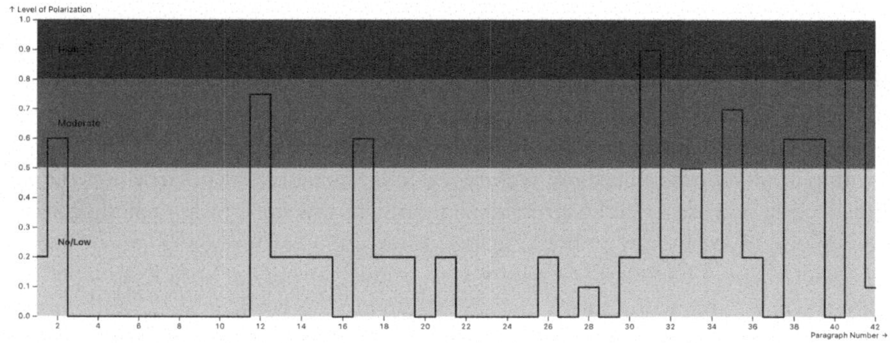

Fig. 7.5 Step chart depicting the evolution of the polarization during a speech of A. Tsipras. The y-axis corresponds to polarization values, from 0 to 1, and colored areas to None/Low (light gray), Moderate (gray) and High (dark gray) levels of polarization sentiments

- A radial dendrogram showing the entities named by each political leader, organized by type.
- A bar plot counting the number of times "will" had been mentioned by each political leader across the campaign period.

7.3 Findings and Insights

Understandably, the volume of data resulting from such an analysis of 171 pre-election speeches is particularly large. A detailed description of the findings would be impossible in the limited scope of a chapter and for this reason in this paper we will focus on some indicative findings that highlight the usefulness of the chosen methodology and the synergy of human expertise and AI tools. We refer interested readers [27] for a more detailed analysis of the findings. Certainly, the first and general advantage is related to the speed of data processing. As we will see below, the speed of processing, the constant and timely updating of our publications, and the visualization charts (see Sect. 7.2.4) that offer a quick aggregate view of the analysis, allowed us to conduct analyses regarding the agenda and the style of each politician and draw preliminary conclusions in a shorter time frame.[8]

The (unexpected) absence of polarization and populism. According to initial expectations at the beginning of the pre-election period, the country would be faced with a particularly polarized political landscape marked, at the same time, by high levels of populism. This estimation, reinforced by the certainty of the two successive elections, leads us to empirically explore this hypothesis by incorporating indica-

[8] To be sure, quantitative and in depth analysis is still needed and generally requires time. Our work does not substitute in depth academic analyses but strives to provide tools to support them.

tors [9, 10] of polarization and populism into the analysis. Eventually, our findings did not confirm these assessments. On the first hand, no extreme levels of polarization or strategic and consistent use of polarizing discursive patterns was observed in the speeches of the political leaders: the average populism and polarization scores over all speeches were 0.2 and 0.1, respectively, on a scale from 0 to 1. Nevertheless, it is important to note that there were occasional instances of polarization during both election periods.[9] On the other hand, populism was not a prominent discursive strategy employed by any political leader. More frequent uses of populist discursive patterns could be found in A. Tsipras' speeches during the first election period, although they remained at fairly low levels (slightly higher than 0.1), while in the second the use of populist rhetoric was almost completely absent.[10]

What (topics) and how (sentiment) politicians talked about. An important advantage of our project was the fact that, by identifying and categorizing the primary topics discussed by political leaders (e.g., Fig. 7.2), we were able to assess their policy priorities. More specifically, our ability to discern which issues were high or low on each party's agenda was greatly enhanced by the use of AI. Comparing the political style and strategy of each leader, our analysis points to pertinent remarks regarding political communication during a campaign period. K. Mitsotakis' discourse demonstrated impressive consistency, characterized by a positive sentiment (as demonstrated in Fig. 7.4) and a dedicated focus on promoting the party's programmatic agenda (more than 85% of his speeches). Moreover, it is worth noting that the thematic agenda around the topics of *healthcare* (9%), *economy* (8%), and *employment* (7%) championed by New Democracy shaped to a great extent the overall debate. This positive and programmatic dimension, which are closely intertwined, saw a slight increase in emphasis during the second election period, likely influenced by the first positive outcome. Finally, Mitsotakis refrained from mentioning his political opponents by name.

Regarding the other political leaders (Fig. 7.3), A. Tsipras and N. Androulakis maintained a distinctively neutral sentiment in their speeches, whereas D. Koutsoumpas and Y. Varoufakis leaned more towards neutrality but occasionally displayed a negative sentiment. In contrast, K. Velopoulos consistently conveyed a negative sentiment in his speeches. During the second election period, A. Tsipras' discourse underwent two notable changes compared to the first. Firstly, there was a significant reduction in populist discursive patterns, as it was described earlier. Secondly, there was a systematic effort to place greater emphasis on the programmatic agenda of the party (from 50% in the first elections to 54% in the second). In terms of the topics discussed, the political debate primarily revolved around the economy. However, in the second period, Tsipras also sought to highlight the importance of healthcare (18%),

---

[9] Notably, D. Koutsoumpas, general secretary of the communist party, recorded the highest escalation in the polarization index during the second election period (average score of 0.24).

[10] A. Tsipras had articulated a characteristically populist discourse in the past, especially during 2012–2015 [28]. Our findings confirm that populism is not an omnipresent feature of a party or a politician. Instead, populist discourse presents fluctuations over time. To properly identify a party as a populist or not, we need to study its official discourse at any given conjuncture.

which became a central focus of his criticism towards New Democracy. Additionally, he addressed issues of accountability, corruption, and democracy. On the other hand, certain important topics such as the environment, culture, energy, agricultural policy, infrastructure, national security, and migration received less attention (less than 5% each for most politicians).

Emerging topics during the two pre-election periods: the case of migration. Finally, as we argued earlier in depth analysis using various methodologies can and should be carried out regarding the two recent elections in Greece. Our project can offer different points of entrance and data for the formulation of research hypotheses, especially in the fields of agenda setting, discourse analysis and framing processes. Using our multi-faceted approach, we managed to identify very soon that the main thematic change in the agenda and the speeches of every politician during the second period was their stance on the migration issue. The devastating sinking of a fishing boat carrying hundreds of refugees and migrants off Pylos on Wednesday, June 14, 2023, [24] brought the issue of migration and refugees to the forefront of public debate. Until the aforementioned tragedy occurred, the issue of migration was either completely absent from the agenda of political leaders or received only passing mentions in their speeches. Of particular interest is how the issue was framed and the connotations associated with it. Indeed, a strikingly contrasting picture emerges when examining the election speeches of the leaders of the two largest parties, New Democracy and SYRIZA. Although migration was not a central theme of K. Mitsotakis' campaign speeches, he mentioned it more frequently than any other political leader. In nearly half of his speeches during the first election campaign, there is at least one mention or passing reference to the immigration issue. These references were present in the second election period, both before and after the deadly shipwreck. However, it cannot be claimed that immigration is a top priority on K. Mitsotakis' agenda, as an overall analysis of his election speeches reveals that migration constitutes only 2% of his programmatic discourse. These references primarily emphasize a specific aspect: border protection. K. Mitsotakis' overall approach to migration in his election speeches revolves around this perspective and the issue was framed as exclusively a matter of national security. On the other hand, the migration issue was completely absent from A. Tsipras' campaign speeches up until the last week of June elections.[11]

7.4 Effectiveness and Accuracy of ChatGPT

As previously outlined, the analysis of the political speech was conducted on a per-paragraph basis. For each paragraph, ChatGPT assigned values to various metrics and categories, which are detailed in Table 7.1. Human annotators made corrections

[11] It is important to underline that only the leaders' election speeches were analyzed (party events, interviews, statements, articles, etc. were not taken into account).

Table 7.1 Accuracy of ChatGPT annotations (with respect to manual annotation of humans) per metric

Task	Most prevalent category (%)	Accuracy (%)
Sentiment	Positive (40%)	94.2
Topics	Elections (26%)	61.7
Polarization	Zero/Low (86%)	87.4
Populism	Zero/Low (96%)	89.8
Criticism or agenda	political agenda (61%)	89.3

to the categories assigned by ChatGPT when they believed that a different category better matched the content of that part of the speech.[12] To assess ChatGPT's accuracy in comparison to human annotations, we calculate it as the *fraction of paragraphs where the ChatGPT annotation remains unchanged by human annotators, divided by the total number of paragraphs.*

The accuracy results, presented as percentages, can be found in Table 7.1. In the ideal scenario, where ChatGPT's annotations align perfectly with those of human annotators, the accuracy would be 100%. Additionally, we include the frequency of the most commonly occurring category for each metric. This frequency represents the accuracy of a "dummy" model and serves as a baseline for comparing the accuracy of ChatGPT's results.

Sentiment: In the task of sentiment detection, ChatGPT attains a notably high accuracy rate of 94.2%, the highest among all tasks. This suggests its robust suitability for sentiment analysis in political speeches, a conclusion aligned with recent findings in the field [1, 31].

Criticism versus political agenda: ChatGPT also exhibits a remarkable ability to differentiate between politicians critiquing their opponents and presenting their own political agenda in nearly 9 out of 10 instances.

Polarization and populism: The accuracy metric for detecting speech containing elements of polarization or populism are high as well. However, it is important to note that in both cases, the majority of paragraphs were annotated -by an expert political scientist- as lacking these characteristics (see "Zero/Low" percentages in Table 7.1). A dummy classifier consistently labeling text as "Zero/Low" for polarization or populism would achieve similar accuracy. These results suggest that ChatGPT may not be suitable for unsupervised detection of such specific political speech concepts. In fact, ChatGPT tends to overestimate the level of polarization/populism in text, providing results more aligned with a broad interpretation of these concepts (as opposed to the specific definitions in political science, as outlined in Sect. 7.3). This overestimation may result from biases or inaccuracies present in the training data used for ChatGPT.

[12] Only for the "sentiment" metric we do not correct, but just annotate as incorrect when ChatGPT places a paragraph into a different category than human judgment.

Topics: ChatGPT assigns a topic from a predefined list of 33 topics to a paragraph, aligning with human annotation in 61.7% of cases. While this accuracy rate may seem relatively low, it is essential to consider that this task is inherently challenging (i.e., multiclass classification with 33 classes), and ChatGPT outperforms a dummy model by nearly threefold (26%). To further validate this, we trained a state-of-the-art NLP model [19] on our dataset and evaluated its accuracy. Even though it was trained on the same dataset and used human labels (an opportunity that was not available during our project's inception), it achieved only slightly higher accuracy than ChatGPT.

7.5 Conclusions

The interdisciplinary nature of our team, composed of individuals with diverse backgrounds within the working group, prompts us to emphasize that, based on our experience with this project, journalism could explore and delve into specific areas, political science could focus on others, and computer science could comment on different, additional aspects. This does not imply that there are no shared findings; instead, it underscores that experts from each field may approach these shared findings differently based on their respective criteria and expertise.

Overall, as expected, the major advantage of integrating AI into this work was the elimination of the initial data processing and analysis speed bottleneck. This comes with additional benefits: First, it allows journalism to provide constantly and timely updated pieces, which can be published shortly after the speeches are delivered. Second, it enhances the journalistic content itself by facilitating timely comparative analysis of the agenda and rhetorical style of each politician. Third, it enables the review of initial assessments early on, shedding light on positions influenced by public sentiment or narrative when they do not align with scientific criteria. This was evident when we discovered that there were low levels of polarization/populism, contrary to what was discussed in the public sphere.

Furthermore, a significant advantage of this type of thematic analysis is that the identified topics in politicians' public speeches allow researchers and the general public to assess their policy priorities based on their entire body of discourse. Alternatively, one can focus on specific speeches to evaluate how these communicated priorities might vary depending on the local audience. Having a continuously updated set of topics that political leaders typically raise can be a valuable resource for revisiting and examining their positions over time and whether their priorities have changed in response to significant events that may influence political rhetoric. This was evident in our project when analyzing their stances on migration issues before and after the tragic sinking incident.

On the other hand, when examining the effectiveness and accuracy rates of ChatGPT, it becomes evident that while it is certainly suitable for specific tasks, such as sentiment analysis, its overall success may not be consistently reliable in all situations. In the event of redoing the topic analysis, we would adopt an alternative methodology. Namely, we would prompt ChatGPT to autonomously discern each paragraph's prevailing theme, followed by classifying these themes into one of our 33 pre-defined topics. In any case, the results obtained from the topic detection task reveal a thought-provoking aspect of using AI in journalism. While ChatGPT can achieve state-of-the-art accuracy, it may still fall short of meeting journalistic standards, particularly when dealing with sensitive subjects like political reporting.

This highlights the necessity for human oversight and underscores the importance and requirement for human-in-the-loop AI processes and tools in journalism. This need for human oversight is also apparent in the process of prompting and providing context, which is aimed at reducing bias and minimizing the inclusion of outdated information in ChatGPT's responses.

Nonetheless, the recent opportunity to train an interactive AI model based on custom data is a promising development for future studies. It also has the potential to shift human involvement from verification to analysis.

References

1. Y. Bang, S. Cahyawijaya, N. Lee, W. Dai, D. Su, B. Wilie, H. Lovenia, Z. Ji, T. Yu, W. Chung, et al., A multitask, multilingual, multimodal evaluation of chatgpt on reasoning, hallucination, and interactivity. arXiv:2302.04023 (2023)
2. C. Beckett, M. Yaseen, *Generating Change: A Global Survey Of What News Organisations Are Doing With Artificial Intelligence* (JournalismAI, LSE, 2023)
3. E. Borges-Rey, Unravelling data journalism: A study of data journalism practice in british newsrooms. Journal. Pract. **10**(7), 833–843 (2016)
4. M. Broussard, N. Diakopoulos, A.L. Guzman, R. Abebe, M. Dupagne, C.H. Chuan, Artificial intelligence and journalism. J. Mass Commun. Quart. **96**(3), 673–695 (2019)
5. H. Cools, N. Diakopoulos, Towards guidelines for guidelines on the use of generative ai in newsrooms (2023), https://generative-ai-newsroom.com/towards-guidelines-for-guidelines-on-the-use-of-generative-ai-in-newsrooms-55b0c2c1d960
6. Data Journalism: Survey – state of data journalism 2022 (2022), https://datajournalism.com/survey/2022/
7. M. Deuze, C. Beckett, Imagination, algorithms and news: Developing ai literacy for journalism. Digit. Journal. **10**(10), 1913–1918 (2022)
8. N. Diakopoulos, The state of ai in media: From hype to reality (2023), https://generative-ai-newsroom.com/the-state-of-ai-in-media-from-hype-to-reality-37b250541752
9. A. Galanopoulos, Detecting polarization in the pre-election political discourse in greece (2023), https://lab.imedd.org/en/polarization-greek-political-speech/, iMEdD Lab
10. A. Galanopoulos, Populism in pre-election political discourse in greece (2023), https://lab.imedd.org/en/populism-in-pre-election-political-discourse-in-greece/, iMEdD Lab
11. F. Gilardi, M. Alizadeh, M. Kubli, Chatgpt outperforms crowd-workers for text-annotation tasks. arXiv preprint arXiv:2303.15056 (2023)
12. A. Hassan, A. Albayari, The usage of artificial intelligence in journalism. Future of Organizations and Work After the 4th Industrial Revolution, p. 175
13. A. Hermida, M.L. Young, Data journalism and the regeneration of news. Routledge (2019)
14. E. Hoes, S. Altay, J. Bermeo, Using chatgpt to fight misinformation: Chatgpt nails 72% of 12,000 verified claims (2023)
15. F. Huang, H. Kwak, J. An, Is chatgpt better than human annotators? potential and limitations of chatgpt in explaining implicit hate speech. arXiv:2302.07736 (2023)
16. iMEdD: A special edition of iMEdD for the general elections in greece (2023), https://lab.imedd.org/en/elections-2023/

17. A.R. KhudaBukhsh, R. Sarkar, M.S. Kamlet, T. Mitchell, We don't speak the same language: Interpreting polarization through machine translation. In: Proc. of the AAAI Conference on Artificial Intelligence. vol. 35, pp. 14893–14901 (2021)
18. K. Kiki, T. Troboukis, A. Galanopoulos, P. Sermpezis, S. Karamanidis, I. Dimitriadis, How we analyze the campaign speeches of political leaders (2023), iMEdD Lab, https://lab.imedd.org/en/pos-analyoume-tis-proeklogikes-omilies-ton-politikon-archigon/
19. Y. Liu, M. Ott, N. Goyal, J. Du, M. Joshi, D. Chen, O. Levy, M. Lewis, L. Zettlemoyer, V. Stoyanov, Roberta: A robustly optimized bert pretraining approach. arXiv:1907.11692 (2019)
20. R.E. McBride, The ethics of data journalism (2016)
21. V. Naxera, V. Kaše, O. Stulík, 'the more populism types you know, the better political scientist you are?' machine-learning based meta-analysis of populism types in the political science literature. Jnl. of Contemporary European Studies (2023)
22. A.L. Opdahl, B. Tessem, D.T. Dang-Nguyen, E. Motta, V. Setty, E. Throndsen, A. Tverberg, C. Trattner, Trustworthy journalism through ai. Data & Knowledge Engineering **146**, 102182 (2023)
23. OpenAI: Chatgpt: Conversational ai service (2023), https://openai.com/blog/chatgpt
24. I. Piper, J. Lee, C. Parker, E. Labropoulou, Tracing a tragedy: How hundreds of migrants drowned on greece's watch (2023), https://www.washingtonpost.com/world/interactive/2023/greece-migrant-boat-coast-guard/
25. S. Rogers, J. Schwabish, D. Bowers, Data journalism in 2017. Google News Lab (2017)
26. M. Sallam, N.A. Salim, B. Ala'a, M. Barakat, D. Fayyad, S. Hallit, H. Harapan, R. Hallit, A. Mahafzah, B. Ala'a, Chatgpt output regarding compulsory vaccination and covid-19 vaccine conspiracy: A descriptive study at the outset of a paradigm shift in online search for information. Cureus **15**(2) (2023)
27. P. Sermpezis, S. Karamanidis, E. Paraschou, I. Dimitriadis, S. Yfantidou, FI. Kouskouveli, T. Troboukis, K. Kiki, A. Galanopoulos, A. Vakali, AgoraSpeech: a multi-annotated comprehensive dataset of political discourse through the lens of humans and AI. arXiv:2501.06265 (2025)
28. Y. Stavrakakis, G. Katsambekis, Left-wing populism in the european periphery: the case of SYRIZA. Journal of political ideologies **19**(2), 119–142 (2014)
29. J. Stray, Making artificial intelligence work for investigative journalism. Digit. Journal. **7**(8), 1076–1097 (2019)
30. B. Zhang, D. Ding, L. Jing, How would stance detection techniques evolve after the launch of chatgpt? arXiv:2212.14548 (2022)
31. Y. Zhu, P. Zhang, E.U. Haq, P. Hui, G. Tyson, Can chatgpt reproduce human-generated labels? a study of social computing tasks. arXiv:2304.10145 (2023)

The manufacturer's authorised representative in the EU is Springer Nature Customer Service Centre GmbH, Europaplatz 3, 69115 Heidelberg, Germany. If you have any concerns regarding our products, please contact ProductSafety@springernature.com

Printed and bound by CPI Group (UK) Ltd, Croydon, CR0 4YY
16/09/2026
02217818-0001